TRADITIONAL AGRICULTURAL AND WATER TECHNOLOGIES OF THE THAR

TRADITIONAL AGRICULTURAL AND WATER TECHNOLOGIES OF THE THAR

BHARAT JHUNJHUNWALA

Publications

KALPAZ PUBLICATIONS

DELHI-110052

Traditional Agricultural and Water Technologies of the Thar

Published in 2005 in India by
Kalpaz Publications,
C-30, Satyawati Nagar, Delhi-110052
E-mail : Kalpaz@hotmail.com

Laser Typesetting at : Shree Vaishnavi Creations, Delhi
Printed at : Mehra Offset Press, Delhi

Contents

Foreword

Palaces and forts, numerous festivals and fairs, colourful folk culture and dress, hospitality and a serene environment with clear blue skies of desertic tract of Rajasthan have a charming effect on the visitors and tourists alike. But for those inhabiting the region, making a living is a big challenge. Barring the fortunate few with access to irrigation from canals or tubewells, the large majority has to depend upon the meager rainfall and the resultant land produce. Surrounded by hope and suspense, from month of June onwards people start looking towards the eastern sky for the first monsoon clouds to appear. When the rains begin everyone rushes to the fields to make most of the situation. A failure of rains thereafter means withered crops as well seasonal vegetation and a year of scarcity and misery. Worst happens when such poor rainfall years come in a row. With hardly any carry over of reserves of food and fodder from previous good years due to increase in human and livestck populations, worries are writ large on how to keep the body and soul together and fend the precious livestock. Large migration of able-bodied men leaving the hapless aged, women and children behind, distress sale or abandonment of livestock, indebtedness and malnutrition are the consequences. For the government it means diversion of huge amount of scarce development funds to drought relief. Though living becomes difficult, hope is still there to look and wait for better times.

The above apart, the challenge of making a living in a harsh and uncertain environment calls for specialised approaches, management practices and lifestyle that entail minimum cash inputs, permit risk aversion yet enable maximisation of productivity of the occasional good year. Mixed farming, i.e. blending arable farming with animal husbandry is an illustrious example. The crop residues and left over vegetation get utilised in rearing animals. The system not only adds to and prolongs the period of otherwise seasonal employment but acts also as a source of ready cash and an asset to tide over difficult periods. Agroforestry, that is conscientious maintenance of multipurpose khejri tree in crop lands, mixing of short and long duration crops as well as their scheduling to cover variation in amount and distribution of rainfall are some of the other practices in this genre. Tankas, khadins, beries and nadis to harvest precious runoff for community use are remarkable mechanisms evolved through human ingenuity. Veterinary care from indigenous medicinal materials, upkeep of herds, and protection of newly born at home and during migration are praise-worthy.

Despite their appropriateness, depth and profoundness, several of these practices and skills have come under severe stress due to greatly increased population pressure and degradation of common property resources. Nadis are getting neglected due to availability of water from tube-well- based water supplies. Yet, the indigenous practices and technologies are epitomes of human endeavour and experience over the generations and several of these are unique to the region. Dr. Bharat Jhunjhunwala under the aegis of Gramin Vikas Vigyan Samiti and with help of its field workers has rendered a pioneering effort in documenting the existing knowledge and practices and in analysing these in terms of their viability and future scope. One may not agree with some of the analysis and interpretations, but the document remains a very useful reference material for all those interested in understanding indigenous know-how and in strengthening its application for a better life in the desert.

Date. 27. 02. 2003

Dr. R. P. Dhir
Former Director
Central Arid Zone Research Institute
Jodhpur (Raj.)

Preface

Indian civilization is today threatened by the onslaught of the West. Of course, the West has its strengths particularly in the areas of technology and economic progress. But the history of the West is really short in the time scale of civilizations. Many insurmountable problems face that civilization today. These include those of environmental degradation, psychological distress and meaninglessness of life. It is not clear at the present moment when these problems will run their course.

In the race for modern technologies, we should not lose sight of our indigenous technological base. It may well be that few decades down the road the Western civilization and the technologies that it has created may be found to be unsustainable. It would be a great loss indeed if we were to lose our indigenous technologies in the interim period. We would then have to reinvent the wheel. Therefore, it is necessary to develop an understanding of our indigenous technologies.

GRAVIS is implementing a project named People's Science for Rural Development in Arid Regions of Thar Desert with the assistance of United Nations Development Programme and Government of India. This report was commissioned by GRAVIS to fulfill the following two mandates of the project:

1. Conducting studies on indigenous techniques of agriculture and soil water management.

2. Documentation of indigenous knowledge and practices.

This study is a small step in that direction. Traditional agricultural technologies have been documented to enable the reader grasp their functioning. It is hoped that similar studies in other sectors and in all parts of the country will help us develop a better understanding and be useful not for ourselves alone but for all people of the world.

Bharat Jhunjhunwala

Acknowledgements

Gramin Vikas Vigyan Samiti (GRAVIS) is a voluntry organisation working in the field of rural development and rural reconstruction in Western parts of Rajasthan since the last 19 years. While the agency has undertaken a number of projects focusing on development of natural resources, judicious management of these resources in a sustainable manner has been a prime concern. We believe that traditional resource management practices that evolved over centuries in tandem with the local ecological balance have much to offer in this regard.

The last three decades have seen a gradual erosion of traditional agriculture parctices in the desert under the onslaught of modern technology. In fact some of these practices may be on the verge of being wiped out. Therefore GRAVIS felt the necessity of undertaking an inquiry into the current existence and frequency of these practices. Hopefully, this will lead to greater awareness in official and enlightened circles about traditional agriculture in Thar.

The documentation has been undertaken by Dr. Bharat Jhunjhunwala, an eminent economist and development consultant based in Delhi. Sh. Pawan, Sh. Sita Ram ji Bhojak, Sh. Shri Kant, Sh. Anada Ram, Sh. Bhikha Ram, Sh. Bhura Ram, Sh. Chaina Ram, Sh. Kana Ram and Sh. Rewat Ram have undertaken the field work for this study. They have put in hard work in this study while bearing other responsibilities of the organisation. Late Sh. Ajay Kumar had coordinated this work in its initial stages. But the Lord has taken away his mortal self from between us. Shri Roshan Lal helped coordinate the study in the final stage. Shri Laxmi Chand Tyagi sat through meetings with the field investigators despite his other pressing engagements. He constantly prodded to look deeper into the issues. He also brought to our attention the existence of many traditional technologies, which may have otherwise been omitted by us in the study. Dr. Prakash

Tyagi was ever helpful in coordinating the visit and making other administrative arrangements. Shri S. K. Kashyap was ever helpful in the overall organisation of the framework of this study by participating in the discussions. Scientists from CAZRI, Sh. L.N. Harsh, Sh. V.K. Manga and Sh. D.Kumar have helped by explaining the underlying concepts of plant breeding and other technologies. The villagers gave us their valuable time and provided information that only they had. GRAVIS is thankful to all of them for the contributions that they have made. Special thanks are due to Sh. Komal Kothari, Sh. Poonam Chand ji Bishnoi and many other senior government officers and scientists who participated in a one-day workshop to comment upon the draft report and gave their suggestions.

In the end, Gravis would also wish to express its gratitude to the United Nations Development Programme and Ministry of Rural Development, Government of India who provided the financial support to make this study possible.

Shashi Tyagi
Secretary
Gramin Vikas Vigyan Samiti

I

Introduction

The *Marusthali* or arid zone of Rajasthan, known in parts as *Shekhawati, Bagad, Jangala* or the Thar Desert, lies in the north-western part of Indian country. The area is characterized by low and erratic rainfall, abundant sunshine, sparseness of vegetation cover, sprawling undulating sandy plains with frequent dunes and widely dispersed settlements.

The mean annual rainfall over the Thar is a little less than 500 mm at the eastern margin and from there it declines to less than100 mm to the west of Jaisalmer. During some years the rainfall is close to normal, few years there is large excess and many years there is subnormal rainfall. In 2000, for example, almost the entire rainfall of the season occurred in the month of July with a total dry period thereafter. This resulted in hardly any food or fodder production.

About seven hundred species of grasses, shrubs and trees grow in the area. The perennial plants amongst these are hardly, tenacious enough to withstand extended droughts, efficient builders of biomass and provide nutritious food and fodder. An excellent example is the khejri tree which is an outcome of the conscience effort to promote and maintain this tree.

Luni River is the only organised drainage system in the Thar. Flow occurs in an occasional year of good rainfall. Several tanks or reservoirs have been built to harness the runoff. Rest of the region has some runnels or drainage lines that emanate from the widely scattered hills or rocky uplands. These are preferred sites for construction of ponds and khadins. Ground water is not only scarce but also saline.

The influx of people into this region followed the invasions from the West starting with Alexander in 325 B.C. Settlements multiplied over the years and also spread towards the more arid parts to the extent that by the turn of the second millennium A.D. even the driest region, namely the Jaisalmer, had not only been settled but also politically organized. Several villages in the desert are 500-1200 years old, with some owning their ancestry to sites not far from their present ones. However, all through the mediaeval and earlier part the twentieth century human population was much less than that at present. The period thereafter has witnessed a phenomenal growth. Thus the population doubled in the 35 years from year 1921 and doubled again following 25 years. The trend has persisted since then.

The rural society in the arid zone is stratified into upper, intermediate and lower castes. The upper and intermediate castes, Jats, Bishnois and Rajputs are engaged in agriculture. There are several other castes with traditional functional roles such as potter, carpenter, etc. Because of reduced requirement of the traditional services. These groups have often taken to agriculture as a sole activity. Professional animal breeders constitute a small part of the society. Shrinkage of area and severe degradation of grazing land is forcing several of these animal breeders to take to agriculture or to migrate to urban centres for alternative jobs.

The general settlement pattern is that of compact settlements, each comprising of about 80 to 500 households. However, due to large increase in number of households and expansion of cultivation, more and more households are making their dwellings close to their agricultural lands in the dhanis.

There has been much expansion of crop cultivation in the recent past. This expansion was considerably aided by the land reforms of the post-independence era. The current pattern of landuse is that two-thirds of arid zone is under cultivation. Most villages have 2-4 per cent of the area under community lands : gauchars (village grazing land) and agor (catchment of the village pond). There are a few orans (lands in name of a deity or a local hero).

Crop husbandry is the major activity. Agriculture is essentially mixed farming with most farmers maintaining a sizable herd of livestock that is sustained on crop residues. Livestock provide a yearlong employment, particularly for the women. Cash from the sale of livestock and the cultivation of commercial crops like guar meet the family needs. Well-off farmers are those with irrigation facilities or those with income from external sources. The number of livestock exceeds human population. The number of livestock has grown from 10 million in 1951 to nearly thirty million in 1997. A perceptible shift in favour of sheep and goat over the period is seen. This recent trend in smaller ruminants is dictated partly by the deterioration of grasslands and partly by marked reduction in demand of to replacement of bullock power by tractors.

Increasing population pressure and appearance of nuclear households has decreased the holding size markedly in the past few decades. The current size of holding does not provide surpluses of fodder and foodgrains that earlier served as reserves for the drought years. Thus a drought year means large scale migration and distress sale of livestock.

Development of irrigation has been a major change in the past few years. The irrigated area constitutes 27 per cent of the gross sown area. Irrigation is undertaken mostly by canals or ground water.

Agroforestry is another important practive evolved by ingenuity of the people. Most agricultural fields have 30 to 50 full-grown *khejri (Prosopis Cineraria)* trees, each of which provides 15-25 kilogram of dry foliage. Scientific studies of the system have established the outstanding merit of this management system. The khejri tree adds to the soil fertility and provides protection against wind erosion but it does not compete with the standing crop for moisture. Because of its deep root system, the tree is able to provide loppings even during the worst of droughts and thus affords some sort of survival insurance. Nummularia shrub *(bordi)* is another plant in this category.

Mixed cropping, i.e., sowing a number of crops in combination is a practice of the area. Short and long duration crops (leguminous crops, cereals and oilseeds) are grown in a mixture so as to cover to the extent possible erraticity in the distribution of rainfall.

Migration of livestock is a major coping mechanism against variation of rainfall. Even farming households hire the service of a volunteer or pastoral group for the purpose. Migration takes place on determined routes and to regions far off from the homestead. Migration of people in search of livelihood to towns and cities is a common feature, the incidence being higher during drought years.

II
Methodology

The study was carried out by the Gramin Vigyan Vikas Samiti (GRAVIS) at its six centres located at Baap, Jelu Gagadi, Kalron, Lawan, Makarana and Shiv in Western Rajasthan.

A checklist of traditional agricultural technologies was made through an in-house discussion with GRAVIS staff. Areas in which the said technology was practiced were marked. Then the study of a particular item was earmarked to three out of the six centres keeping in mind the geographical distribution. Most studies have been undertaken in two or more centres.

A semi-structured format was given to the field investigators to fill in with respect to each technology. The reports of the field investigators were discussed in the group and supplementary studies were undertaken where necessary. The following instructions were given in regard to the studies:

- Discuss with two persons each using traditional and modern technologies while filling in the report.
- Study in particular fields which may have been damaged by the use of modern technologies.
- If variation is found in the use of technologies in different villages then more than one separate report may be made.
- Use local terminologies in the reports.
- Listen and record what the farmer tells. Speak less and listen more.
- Give special attention to cultural dimensions of the technologies such as related folk songs and religious festivals. Record their impact on

children and women.

- Record the details of the traditional technologies as per memory of the respondent even if it is not found to be in use at the present time.

Field visits were undertaken by the lead investigators in each of the field areas to supplement the efforts of the investigators.

The field investigators were given a camera and requested to take photographs of the technologies in operation.

Mid-way during the study a doubt arose regarding the extent of use of traditional or modern technologies. It was found that guess estimates varied substantially. It was therefore decided to request the investigators to make a quick survey of ten families in their area to ascertain the use of the two technologies. The purpose of this survey was to get a quick assessment of the present state. The surveys have not followed scientific sampling methods. They should only be taken as a superficial overview of the use of traditional technologies to give the reader a bird's eyeview of the situation.

III

Traditional Agricultural System of Western Rajasthan

The unique feature of the traditional agricultural technologies of Western Rajasthan is that grazing land, cropland, livestock and the rural homestead have been fully integrated with one another in an organic way. Each of these systems was dependent upon the others. This has been shown in the table below.

Connection of	*Connection with*	*Items provided*	*Important Technologies that are affected*
Grazing Land	Small animals	Grazing	*Adav*, Common Grazing Lands
	Large animals	Grazing	
	Cropland	Biomass for manure	*Jhoor, Kanabandi, Baad*
Small animals	Homestead	Meat	
Large animals	Cropland	Draught power and manure	Ploughing
	Homestead	Milk, fuel	*Thepadi*
Cropland	Large animals	Fodder	Silvi-agriculture
	Homestead	Grains	

The rural homestead received livestock from goat and sheep, milk from cows and grains from cropland. There was a close connection between cropland and the cow and bullock. The cropland provided fodder. Cows and bullocks provided draught power and manure. The large animals were the axis around which all the activities of the system revolved. They were connected with both the grazing lands and croplands. They were indispensable for crop cultivation because they provided both the manure and draught power. This manure they partly generated from the food they got through grazing. Grazing lands were indispensable because they provided the manure through the large animals.

The advent of tractors and chemical fertilizers has fundamentally changed the situation. Large animals are losing their importance since draught power is available through tractors. The buffalo is replacing the cow. The crucial difference between the two is that cows are partly dependent on grazing from common lands while buffalo is a sedentary animal. The less importance of the bullock has led to a parallel shift from cow to buffalo. Previously the common grazing lands were not allowed to be overgrazed by small animals in order to provide grazing for the cow. The common grazing lands are not being much protected now because large animals have become less important. The common grazing lands also provided biomass for fencing (*Baad*) and manure (*Jhoor* and *Kanabandi*). The denudation of common grazing lands has reduced this availability of biomass. But the urban system has provided chemical fertilizers to partly make up for the loss. However, the use of chemical fertilizer may have negative consequences for the soil which has less moisture and less organic matter.

The grazing lands were not encroached earlier because they provided grazing for large animals. Encroachment would have reduced this grazing and hit at the availability of draught power for crop cultivation. The introduction of tractors has made it possible to cultivate larger areas of cropland without having to depend on grazing by large animals. That has led to less interest among the people in the protection of common grazing lands.

The modern system of agriculture that is gradually replacing this integrated system is heavily linked with the urban system. The city now provides three crucial inputs to the rural homestead:

Chemical fertilizer to replace manure obtained from grazing.

Petroleum products to provide draught power and replace bullocks.

Wage income to the people.

The buffalo is often maintained in the city itself. Being a sedentary animal it can be kept in small spaces. The need to convert fodder into milk in the rural area, therefore, has also reduced. These changes have been shown in Fig. 2. It will be seen that the modern system is much more urban-centred. The integration of cropland, large animals and grazing lands has been broken. Both large animals and grazing lands have been rendered less important to cropland cultivation. The main connection of the modern system with the village is through croplands.

The technologies relating to *Adav*, Common Grazing Lands, *Jhoor*, *Kanabandi, Baad,* Ploughing, *Thepadi* and Silvi-agriculture are consequently under stress. This transformation, however, does not mean that the modern system is sustainable. The availability of fossil feul is limited. The extraction of fossil water is fast depleting the ground water table. There are some reports that the long term fertility of the soil is declining because of less organic manure being applied.

Other traditional technologies such as mixed cropping, *Adav* and *Nadi* are compatible with the transformation outlined above. The challenge before us is to recreate another sustainable agricultural system by modifying and reviving these traditional technologies while benefiting from the recent technological advances mentioned above. The danger we face is that we may lose slight of the more durable traditional technologies in the exuberance of modern advances which may then turn out to be unsustainable. Therefore, there is a need to appreciate the inherent strengths of traditional technologies while recognizing their limitations.

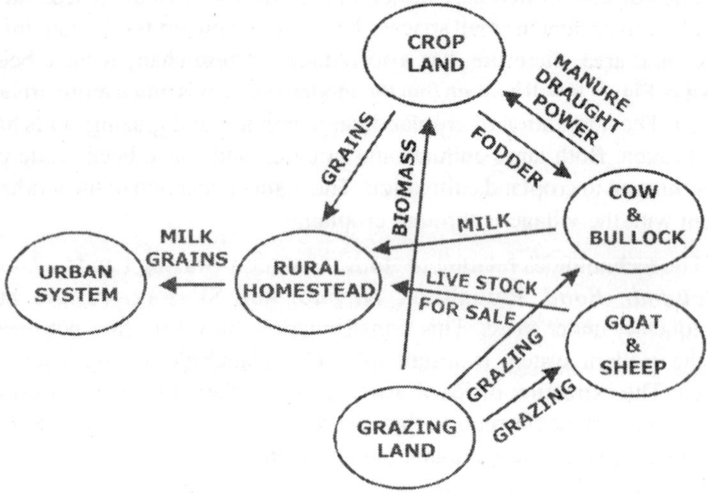

Fig. 1 : Traditional System.

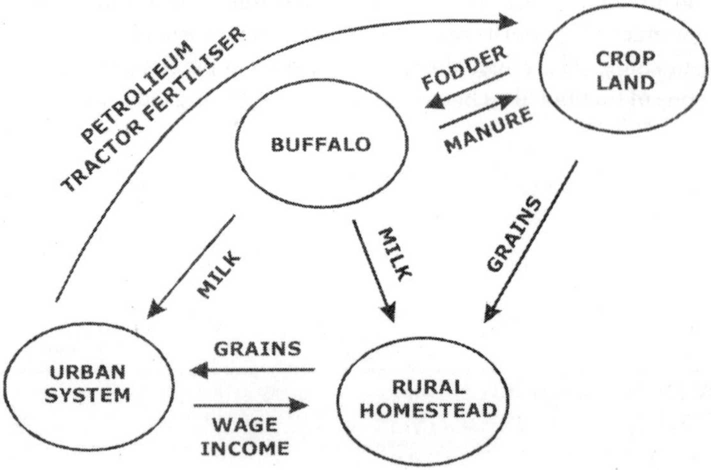

Fig. 2 : Modern System.

IV
Overview

The 50-odd traditional agricultural technologies studied as a part of this assignment can be classified as follows:

Stable: These technologies are holding their place and are not threatened. They are not dependent on subsidies. Nothing much needs to be done with regard to them. They may mainly be watched to take notice of new threats and to respond proactively.

Area	*Stable*	*Potentially Viable*	*Declining*	*Total*
Pre-sowing Agriculture	3	2	2	7
Crop Management	2	4	Nil	6
Harvest and Post-Harvest	2	2	1	5
Trees and Grasses	2	3	2	7
Field Management	Nil	3	1	4
Livestock Management	3	1	2	6
Festivals	Nil	3	Nil	3
Water Management	1	4	1	6
Total	13	22	9	44

Potentially Viable: These are weakening or threatened technologies but there is a possibility of reviving them with some effort.

Declining: These technologies are being practised but the pressures are intense and in the assessment of the author it is less likely that they can be revived. They have been rendered unviable due to modern technological developments. In some cases, their revival is not desirable because their impact on human life is not as beneficial as the modern alternatives. It appears that they may have to be abandoned.

A bird's eyeview of the technologies can be obtained from the table below.

The programme for intervention should primarily be focused on uncertain technologies because the gains of intervention are likely to be greatest here. Detailed suggestions in this respect have been given at the end of individual reports of each of the technologies. The main suggestions are summarized in the next chapter.

STABLE TECHNOLOGIES

Resowing (Bhanj): Resowing will remain an essential feature of dry land agriculture as it increases yield with nominal expenditure.

Mixed Cropping: Farmers are ever adopting various new combination of dry crops depending upon the situation. This is likely to continue because of the need to spread the risk—as an insurance against erratic rainfall, destruction by cattle and at times spread of diseases.

Weeding (Nidaan): *Nidaan* is stable for the *Kharif* crops of millets but is threatened in the Rabi crop of Rayada owing to the need to do it repeatedly and growth of insects that decrease the value of the weeds as fodder. It may be possible to revive it in Rayada if the harmful effects of weedicides are documented and publicized.

Protection from Frost by Watering (Pala): There is no alternative to watering for protection from frost in the case of winter crops like mustard and chili.

Protection from Frost by Castor Plants (Pala): Castor plants are an inexpensive method of protecting chilies from frost.

Harvesting: There is no modern alternative hence this technology is stable.

Food Grain Storage (Kinhara): The traditional technology is both cheaper as well as safer than modern steel containers.

Tree Lopping (Loonkh Katana): Lopping will continue to be done for better re-growth of trees. The choice is between easy- and hard lopping. This can be studied.

Protecting Small Trees (Jhopa): The technology is well entrenched.

Livestock Nursery (Tutadia): Farmers will continue to build *Tutadia*. The modern alternative of stone and cement is also being used.

Fodder Shredding (Kutar): The tractor shredder cuts even hard parts of stalk and is, therefore, more efficient. The technology of shredding will continue albeit with tractors.

Open Fodder Storage (Kalar): Storage of fodder by putting up a *Kalar* will continue because it is cost effective.

Storage of Drinking Water (Tanka): The people deserve access to piped drinking water. At the same time water security requires that roof top harvesting be promoted. Tanka would be an addition to roof top water harvesting.

POTENTIALLY VIABLE TECHNOLOGIES

Traditional Crop Varieties (Desi): Farmers select the variety largely on economic consideration of yields and profits though factors like water availability, type of land used, fodder requirements of livestock, household needs and availability of reliable seed are also considered. Some traditional varieties are under pressure from hybrid ones.

Land Preparation (Sood, Jhoor and Kanabandi): The practice of *Sood* is technically sound and stable. Technologies of *Jhoor* and *Kanabandi* convert waste biomass into manure. These are under pressure mainly because the availability of biomass from grazing lands is declining.

Cultivation in Bund Bed (Khadin): Cultivation in *Khadin* leads to higher income for the farmer and is likely to continue as long as *Khadin* themselves continue to be made.

Insecticide: There is a range of traditional insecticides—from stable to dead. As compared to modern insecticides, they are cheap but more labour-intensive and less effective.

Insecticide (Ash in Jira): The technology is used widely but is threatened by reportedly more effective chemical insecticides. It can be improved.

Protection from Birds: Attack on crops by birds is a perpetual problem. Some technologies may decline but others can be modified.

Seed Selection: The desi varieties can become viable if their yields are improved by seed selection. They are more resistant to diseases and drought hence it is desirable to do so.

Seed Storage: The farmer will continue to save his seed. The modern pucca houses have reduced the threat to seeds from open storage in sacks. Therefore this method may gain greater.popularity.

Silvi-Agriculture: Trees are important for ecological as well as soil stability. But the farmers are not inclined to grow trees within their fields because they eat into valuable cropland, attract birds and draw upon the earth's nutrients. The return to trees should be increased. They may be planted on field lands in a big way.

Storage of Wood (Pindara): Use of wood as fuel has severe ecological implications in view of declining availability of biomass and forests. On the other hand, blind reliance on fossil fuels like kerosene and LPG is also troublesome because of the dependence it builds on imports. The correct direction would be to move towards a sustainable fuel policy based on renewable energy. This would imply cultivation of biomass at the farm level. The technology of Pindara has to be dovetailed into that larger strategy.

Tree Produce (Ber, Kair and Kumatiya): The economic value of the produce is negligible. These trees are becoming extinct. Technology of picking their produce will decline as these trees become extinct.

Fencing (Baad): The traditional baad is holding its position. However, it will come under increasing pressure as larger number of rural people move to cities or engage in non-agricultural vocations. The availability of wage employment throughout the year has meant that the previously 'lean' period of summer, during which the opportunity cost of labour is small, will no longer provide time for making the traditional baad. A pro-active strategy, therefore, would be to reduce the labour involved and increase the returns from the traditional *baad.*

Common Grazing Lands (Oran & Gochar): The best people-friendly use of common lands has to be studied afresh in view of the change from draught to milch animals as the main large livestock.

Land on Rent for Grazing (Bhelwad): This is an ecology- and people-friendly technology. There will always be persons who have less-than-optimal and more-than-optimal number of livestock. This technology helps match land to livestock. No alternative to this technology is available. It is being unnecessarily restricted due to land laws.

Shelter for Livestock (Taap): Livestock will continue to need protection from the weather gods. Traditional and modern technologies should be combined to provide both comfort and protection.

Jawari: This may be a good practice for pre-testing the seed. Some improvisation can revive this technology.

Meteorological Predictions (Sugan) : It is acknowledged that animals have better ability to sense the climatic changes. This knowledge can and must be harnessed.

Hindola: Traditional cultural practices were rooted in the life cycle. The Hindola must have been serving some purpose, the memory of which we have lost. Let us rediscover it.

Bund Construction (Khadin): The technology is under pressure. The making of a khadin through strengthening of the *bund* can be promoted. The technology is fundamentally useful. Design improvements can be made to stangthen it.

Johad: The technology is fundamentally sound. It has been rendered dysfunctional because of the decline of grazing by cows. It has to be adjusted to buffalo requirements.

Village Tank (Nadi): This technology is likely to be used where water from PHED is not available.

Drawing Water from the Tank Bed (Beri) : This technology is a subsidiary technology of *Nadi*. It will grow or die along with the *Nadi*.

DECLIMIMG TECHONOLOGIES

Pre-sowing seed treatment (Gojala) : The efficacy of the technology is unclear.

Ploughing: The time taken by draught animals is much higher than that by tractors. The improvement in breed is unlikely to reverse the situation of decreasing use of draught animals because the gap is very large. This technology is very important because it determines the nature of the homestead. Draught animals bring with them a host of technologies such as drawing water from the *Nadi* bed by *Kalsi*. These second-level technologies have no chance once the draught animal is gone. The use of fodder, however, is likely to shift in favour of milch animals which may not be. sold. Ploughing using draught animals has been classified in the category 'Declining' because the modern alternative involves less drudgery and enables the cultivation of large areas leading to food security. The long-term issue of declining availability of fossil fuels does auger for putting greater effort in the revival of use of draught animals for ploughing but as economist Keynes said, " In the long run we are all dead." It is possible that human ingenuity will invent some technology like solar power to overcome the limitation of fossil fuels.

Threshing (Lata): Mechanical threshing is very fast. The loss to the farmer is immense if the ripe crop is lost due to delay in threshing. The cost of mechanical threshing is also cheaper. For threshing small grains, however, use of *Lata* works out to be the better option.

Harvest and Storage of Ber Stalks (Pai): Ber shrubs grow profusely in the area. They provide biomass for various works such as *Baad* and *Taap.* Their stalks are good material for fencing. But these shrubs are dying because of *tawi* cultivation.

Wild Grasses (Kheep and Sinia): The use of these wild grasses for making houses and ropes is labour intensive.Though use of grasses is cheaper, cement and mortar (for houses) and plastic (for ropes) have benefits like durability.

Fallow Lands (Adav) : Soil can be provided rest by either leaving the field fallow or by turning it with *tawi.* The latter can be combined with crop agriculture. Therefore, *Adav* may decline.

Storage of Dung Cakes (Thepadi) : This traditional technology should be discouraged as it increases the workload for women without having any special significance and produces a lot of smoke. Moreover the dung from which *Thepadi* is made can be profitably used as organic manure in fields. It belongs to a period when other environment-friendly sources of fuel were not available.

Migration of Livestock (Palayan): There is a major shift from cows and bullocks to (1) buffalo (for the use of crop fodder) and (2) sheep-goat (for the use of grazing). Of these, only sheep and goat are made to migrate. The number of these small animals would reduce in proportion to the decline of common grazing lands. Therefore, this practice is likely to reduce in the times to come.

Transport of Drinking Water (Kalsi): The cost of transporting by tractors is very low hence this practice is likely to decline.

V
Key Suggestions

The basic transition in traditional agricultural and water technologies that is taking place is as follows.

- Grazing lands, *cows, bullocks and camels* are declining.
- Crop cultivation will continue though with more use of mechanized implements and less of animal power.
- Trees would have to be proactively cultivated.
- Common lands would have to be reoriented towards fruit and other products.
- Water management technologies will continue and can be improved.

This transition will involve reinvention of some technologies. For example, ploughing by animals will decline but new technologies of ploughing are developing. A large number of suggestions have been given in this regard in this report. In this section some important suggestions have been mentioned which, in the assessment of the author, hold the key to re-establishment of sustainable agriculture in the areas. These suggestions have been given not with the objective of conserving existing technologies but with the objective of establishing a new sustainable agricultural pattern while building on traditional technologies.

The suggestions have been classified as pertaining to the following :

Communication: There appears to exist a gap between the known technology and people's knowledge of the same. A programme of communicating the correct situation to the people may be undertaken.

Government: A change in government policy would help in strengthening the traditional technology.

Study: There is a lack of information. We do not know much about the ground reality. A detailed study may be undertaken to find the correct situation. Follow-up can be planned only after such a study.

Technology: The problem is known and the potential of solving it by technological innovations is also visible. These technologies may be developed. This requires fieldwork.

COMMUNICATION

Land Preparation: Farmers have to be made aware that *Jhoor* should be undertaken over a wider area so that the danger from excessive soil heat due to low rains is minimized.

Seed selection: Fields may be selected for seed production instead of pods. A social system may be created for enabling exchange of seeds of the best fields between different villages.

Silvi-agriculture: Find crop-friendly species like Khejri and Bordi which can be planted. Bajra grows well under these two trees. Planting of such crop-friendly tree may be possible. Farmers are under the wrong impression that Khejri only grow spontaneously through the spread of rootstock and cannot be planted. Scientists at CAZRI, however, indicated that the spontaneous growth of Khejri plants occurs not through rootstock but through seeds. The seed of Khejri can be fed to the goats and the dropping containing softened seeds can be sown in the fields. This technology has to be propagated.

Common Grazing Lands (Gochar and Oran) : A social consensus along the lines of Joint Forest Management should be built on the proper maintenance and use of *Oran* and *Gochar*. The relationship of panchayat elections with *Gochar* and *Oran* is tricky. Panchayats are valuable in themselves as a measure of promotion of democracy. But democracy itself works only if there is a non-political agency to define the public good and spread its awareness among the people. This agency should be created.

Jawari : Farmers must be educated about the significance of *Jawari* as a process of pre-testing a variety of seeds whether bought or home-grown.

Tanka : Water harvesting from rooftops should be added to the ***Tanka*** construction. Previously pucca houses were few. Perhaps pipes were expensive. Thus rooftop water harvesting was not much practiced except in areas like Shekhavati where ground water was brackish and unfit for potable use. Now this dimension can be added to the *Tanka* construction. Demonstration can be made in villages and publicity material can be distributed. This will save valuable land from being used as catchment area.

Nadi: Linking the rules of *Nadi* to religion and festivals can increase the participation of people.

GOVERNMENT

Traditional Crop Varieties: Make available reliable *Desi* seed in packed form.

Land Preparation: One major problem is the availability of the biomass. Previously farmers would bring biomass from grazing lands. Now the grazing lands have mostly been cultivated and this source has dried up. Biomass is only available from wild plants and shrubs that grow in the fields. This is very limited in quantity. Then there are demands on the biomass for use as domestic fuel and for making *bada* for livestock. These leave less biomass for *Kanabandi* and *Sood*. Possible action in this regard would be to encourage the use of alternate domestic fuel such as solar energy or LPG gas.

Gochar and Oran: The legal status of the *Oran* in land laws should be clarified. It may be put at par with *Gochar* to prevent its allotment for private purposes.

Nadi: A method has to be found to enforce the rules for the use of *Nadi*. The Panchayat elections have disabled the traditional consensus-based approach. A window has to be created within the rules of Panchayat to manage local community resources collectively.

STUDY AND RESEARCH

Traditional Crop Varieties: Undertake a study of the impact of *Desi*- and Sankar varieties on health of human beings and livestock.

Insecticide : Undertake comparative field trials of traditional and modern insecticides to study particularly development of immunity among the insects and negative environmental impact.

Food Grain Storage : A comparative study may be made of benefits and damage during storage of foodgrains in *Kinhara* and steel containers.

Seed Storage : A scientific study may be carried out on use of different storage technologies (earthen pots, sacks, etc.) for different types of seeds and the results disseminated among the farmers.

Gochar and Oran : The basic purpose of *Gochar* and *Oran* is to secure public good—in this instance, through making available common grazing land. It has to be, therefore, examined whether grazing land is the best method of securing public good in the present circumstances. It appears that now grazing is a low-priority need in view of the decline of the use of bullocks, which has done away with the need to maintain a large herd of cows. What need of the people can be most served by the common land should, therefore, be examined afresh. Alternative people-friendly uses of common lands should be promoted. These could be used for fruit plantations or water catchment. The suitability of fruit trees of Amla, and Lasoda (Goonda) may be examined in this respect.

The loss of *Oran* is related to the decline of religious sentiments. A meeting may be organized of the priests of *Oran*-dedicated deities. Stories of 'curse' of the past and present should be documented and awareness generated. That will help instill fear among the people and restrain them from cutting trees.

Sugan : A comparative study of *Sugan* and meteorological predictions may be made. The omens used by the Sugani in making the predictions may be noted and tallied with the results. It is necessary to develop this technology as a science. It would be useful to make a survey of the *Sugan* before the monsoons and then compare the actual results with their predictions.

Tanka: It will be advisable to have water samples from *Tanka* checked for suitability as potable water. It is possible that some diseases may be occurring due to use of water from *Tanka*. If so, measures for disinfecting the *Tanka* should be developed and awareness of the same may be spread.

DEVELOPMENT OF TECHNOLOGIES

Traditional Crop Varieties: Undertake a breed selection programme to select and spread desi high-yielding varieties which are in tune with local agro-climatic conditions.

Land Preparation: Jhoor can be done with *kadab*—the stalk of Bajra and other crops. Paddy straw is being regularly ploughed into the fields in Punjab. The leftover biomass of certain crops like Moong and Moth is not used as fodder anyway. A comparative study must be done to ascertain returns from the biomass by its use as fodder and as Jhoor.

The biomass could be chopped in a thresher instead of manually. This may make it possible to chop into smaller pieces, which would increase productivity by more even spreading.

Ploughing: Breed improvement. Try to find or develop a breed of cows or buffaloes which double up as draught animal for short periods.

Resowing (Bhanj) : New types of cultivators with special spikes may be devised for use with tractors so that the blade of the plough attached to the tractor can be steered to save existing plants.

Mixed Cropping : It may be useful to document the benefit of mixed cropping and submit a report to the agricultural universities or organize a seminar to encourage them to undertake plant breeding keeping in mind the synergy of different crops for mixed cropping.

Insecticide : A programme synthesising traditional and modern insecticide treatments many be evolved and tried out.

Seed selection: Developing and disseminating better methods of seed selection can promote the technology. Farmers do not know which of the traits such as a taller plant, bigger size, colour, shine or shape of the grain, size of the pod, etc. are important. A study can be undertaken to select seeds according to these different parameters and the results can be disseminated.

Discussion may be held with seed scientists to determine the optimum level of climate shock that is beneficial for the crops. Farmers may be educated about choosing the area and farmer with whom seed may be exchanged.

Silvi-agriculture: Undertake research on the optimal spacing between trees and between lines of these trees. Planting in lines can enable cultivation by tractors. Farmers have to be made aware that it would no longer be adequate to rely on the spontaneous growth of these trees because they obstruct cultivation by tractor. The need to consciously plant these trees has to be communicated.

Find trees that give high value returns so as to make them more economical in comparison to crop cultivation.

Baad: Study techniques of establishment of *Khejri* tree on the *Baad*. It is clear that trees provide an ideal solution for baad, especially in combination with barbed wire but they occupy considerable land area—about 10 to 15 feet land along the *Baad* is rendered unfit for cultivation. The technology of plantation of *Khejri* on *Baad* should be developed. The main hurdle is giving water to the small saplings after transplantation. Experiments have to be done and the cost worked out to see if the cost of giving water can be recovered.

Study suitability of other species of trees and shrubs for planting on the *Baad.* Other tree species have to be studied for their suitability for silvi-agricultural cultivation. Trees like Kheri and shrubs like Moonja have little synergy with agriculture. Ratanjyoti—a shrub that grows in Southern Rajasthan—is suitable for planting on Baad and it also provides oilseeds, which fetch a good income. Some varieties of agave cactus are known to provide raw material for the production of alcohol and other chemicals. Thor—the cactus that is used in Western Rajasthan—does not provide such commercially valuable products. Experiments can be undertaken with other species of trees and shrubs that simultaneously provide protection and other commercially valuable products.

Khadin: The design of *Khadin* must be improved to minimize the loss due to breach of the *bund.* A structural engineer, as well as a team of village technicians, may be asked to look into the ideal design and dimensions.

The construction of *Khadin* by strengthening the *Baad* by more systematic *Kanabandi* should be studied and propagated. It is a low-cost technology. The topographical conditions (slope, wind speed and direction, soil type, etc.), which make it possible to construct such *Khadin*, may be enumerated and propagated.

VI

Crop Planning and Sowing

1

Crop Planning and Sowing Traditional Crop Varieties (Desi Varieties)

Category: Potentially Viable

Hybrid varieties of *bajra* are cultivated on irrigated lands in selected tracts such as those of Nagaur. However, composite varieties are grown on dry lands. Crossing native varieties with unknown hybrid varieties has made these composite varieties. Two different types of composite varieties are grown in the area—those that have a stronger influence of hybrid varieties and those that have a stronger influence of native varieties. In local parlance the former is known as *Sankar* while the latter is known as *Desi* though, in reality, both are composites. This study is based, in the main, on people's knowledge; thus the terms *Sankar* and *Desi* have been used for the two types of composites.

When are the varieties used ?

Both varieties are sown at the same time—at the onset of rains. But the *Desi* variety takes up to four months to mature in comparison with two to three months for *Sankar* varieties.

Where is sowing done?

The farmer sows these seeds in his field.

How is it done?

The seed is sown.

Cost

Nominally, the cost of *Sankar* seed is higher as it is more often purchased from the market. But this is only a small part of the cost of cultivation. The main issue is that of yield. The *Desi* varieties give higher yield when there is low rainfall while *Sankar* varieties give higher yield when there is high rainfall.

Effect

The basic differences between the two varieties are as follows. This applies to most crops.

Comparison of Desi and Sankar Varieties

Sl. No.	*Desi Variety*	*Sankar Variety*
1.	Fertilizer is not needed or needed in less quantity	It is necessary to add chemical Fertilizer.
2.	Its seed is available locally and free of cost.	Its seed has to be bought from the market.
3.	It can grow in less rain or even no rain.	It cannot grow when there is less or no rain.
4.	The *chapatti* of *Desi* grain is good and more nutritious. water has to be consumed while eating chapatti of *Desi* grain. This indicates its greater density and strength.	The chapatti of Sankar grain is more nutritious and tasty.
5.	More cattle feed is obtained from *Desi* crop.	Cattle feed obtained from Sankar Variety is less.
6.	It is resistant to disease and is able to withstand natural calamities better.	It is prone to diseases and has less capacity to withstand natural calamities.
7.	Price of desi grain is higher than *sankar* variety.	Price is lower.
8.	More effort is required for crushing Desi grain into flour	Price is lower.

Desi seed is almost universally believed to be more nutritious. Many farmers grow small amounts of *Desi* varieties for their own consumption while growing *Sankar* varieties for the market. There is no evidence for *Desi* varieties being more nutritious but it is cited that earlier when women used to make *Khichadi*—a type of porridge—they used to tie cloth on their hands to hold the pot steady because the *Khichadi* used to burst like popcorn. This was taken to be an indicator of the 'strength' of the grain.

Other factors relating to taste as narrated by Meera Kanwar and Amka of village Kalyan Singh Sid of Phalodi are given in the table below:

Serial No.	*Desi Variety*	*Sankar Variety*
1.	It is hard.	It is soft.
2.	One feels satisfied by eating less food.	More has to be eaten.
3.	Eating *Desi* grain makes the body strong.	*Sankar* grain makes the body weak.
4.	Body becomes resistant to diseases.	Eating *Sankar* seed makes the body more disease prone.
5.	Eating *Desi* grain makes one live long. Earlier people used to to live up to the age of 100.	Longevity is reduced. Now peo-people live up to the age of 55 to 60 only.
6.	*Desi* grain improves eyesight.	Eyesight becomes weak.
7.	More strength is required to grind it.	*Sankar* grain can be easily gro-und into flour.
8.	*Desi* grain is sweeter.	*Sankar* grain is tasteless.
9.	Eating Desi seed makes the will power strong.	Eating Sankar grain makes one's will-power weak.
10.	One needs to drink more water when *Desi* grain is eaten.	One needs to drink less water when this grain is eaten.
11.	Eating *Desi* grain makes menst-ruation in women regular.	Eating *Sankar* grain makes men struation in women irregular.

A survey in three villages in different areas brought out that a majority of women continued to prefer the use of *Desi* varieties. At the same time 11 out of 28 persons were using *Sankar* varieties. One may presume that the use of *Sankar* varieties has been increasing. Therefore, the situation appears to be evenly poised at the present.

The main tradeoff lies between low costs of hybrid grains against many other benefits of *Desi* grains.One important observation made by Guman Singh of Jelu was that *Sankar* varieties gradually become like *Desi* ones—in three years the grain becomes sweet and the yield less. Perhaps the agro-climatic conditions have made the *Desi* varieties what they are today. This implies that *Sankar* varieties, which gradually acquire *desi* characteristics, are 'unstable'. It can be concluded that they are not *in tune* with the local natural conditions. In growing such *Sankar* varieties man is working against the design of nature and is likely to trap himself in

Survey

Name of Technique	*Number of respondents using the variety*	*Reasons*
Traditional: Use *Desi* food grain	17	*Desi* food grain is more nutritious. Desi food grain is tasty because organic manure is used. *Desi* grain is sweet.*Desi* grain gives more strength. Women have to tie cloth on their hand to hold the vessel in which *Khichadi* from *desi* grains is being cooked because it jumps. There is less disease and illness.The chapatti of *Desi* grain is soft. The flour of *Desi* grain does not get spoilt for a month.Eating *Desi* seed increases memory. In the past even old persons recalled year old events.
Modern: Use hybrid food grain	11	Hybrid food grain cheaper. It takes less time to clean hybrid seeds for cooking, as the seed is thicker.

more difficulties. The ill-effects on health and soil may be substantial in the long run.

The price difference in *Desi* and *Sankar* varieties in October 2001 were as follows:

It will be seen that the farmers are mobile between *Desi* and Sankar varieties of different crops.

Mularam of Kinsaria cultivated *Desi* and *Sankar* crops as follows:

Bajra: He had grown both *Desi* and *Sankar* varieties. *Desi* variety had grown tall, and provided good amount of fodder and intersowing of til had better done. It provided some yield but the yield was one-half of what had been expected. The *Sankar* variety was mainly monocropped and only sparse intersowing of til was done. It had given good yield but the til had given no yield whatsoever.

Chola: He had grown both *Desi* and *Sankar* varieties. Desi had big vegetation growth but few pods. Sankar had less growth but more pods. Desi pods were sweet while Sankar pods were plain flat in taste.

Price of Desi and Sankar Varieties (Rs. per quintal)

Crop	*Desi*	*Sankar*	*Per cent cultivation of Desi crop as estimated by grain dealers*	*Reason*
Bajra	275	260	5-10%	Yield higher in *Sankar.*
Chola	1150	1700	1%	*Sankar* has bigger grains.
Moong	1900	1900	15%	*Desi* is less sensitive to rainfall
Moth	1550	Na	100%	No *Sankar* variety available
Guwar	800	Na	100%	No *Sankar* variety available

Moong : Only Sankar variety grown. Good crop was expected.

Moth: Only desi variety grown. Diseases affected even this.

It is clear from the above that the farmers are quite free in choosing *Desi* and *Sankar* varieties according to the peculiarities of a crop.

At Jelu Gagadi some estimated that 75 per cent area was under *Desi* varieties while others estimated that 75 per cent was under *Sankar* varieties.

The main factors that influence the selection of variety are as follows:

Water : Sankar is grown where water is available and *Desi* where it is not or uncertain.

Price and Yield : The price and yield differential between *Desi* and *Sankar* influences their use.

Land: *Desi* is grown on uplands while *Sankar* is grown in the basins, as the moisture is more in the latter area.

Livestock: Farmers having livestock grow *Desi* because it gives more fodder.

Household requirement: Some *Desi* grains are grown for self-consumption because they are sweeter.

ANALYSIS

Availability of seed: Seeds of *Sankar* varieties are purchased from the market more frequently. The availability of 'reliable' seed influences the decision. Some farmers were confident of using *Sankar* seeds available from CAZRI alone.

Survey at Village Chandeliyaa Ki Daanni, Nagour

Name of Technique	*No. of respondents Using the technique*	*Reasons*
Traditional: *Desi* seed	3	It is cheap and tasty. One need not buy the seed. Fodder grown is more in quantity. *Desi* crop is disease-resistant. It needs less fertilizer.
Modern: *Sankar* seed	3	*Desi* seed gives less produce while *Sankar* gives more produce. The crop ripens fast, i.e., in a period of 60 days.

SUGGESTIONS

- Undertake a breed selection programme to select and spread *Desi* high-yielding varieties, which are in tune with local agro-climatic conditions.
- Make available reliable *Desi* seed in packed form.
- Undertake a study of the impact of *Desi* - and *Sankar* varieties on health of human beings and livestock.

Desi bajra

Sankar bajra

Only *desi* moth is cultivated

Desi chola on left has good growth but gives poor yield. Sankar chola on right gives good yield

Only improved moong is cultivated

2

Pre-Sowing Seed Treatment (Gojala)

Category : Declining

Seed treatment is done to make the crop resistant to insects and diseases.

When is it used?

Seed treatment is done before sowing the seed in the field.

Where is it done?

It is done at home.

How is it done?

Seed of wheat is soaked in water in the evening. The water is drained in the morning. Then about 1 litre *Gojala* (cow's urine) is added to 10 kg of seed.

The seed is spread in the courtyard and ash sprinkled over it. Then *Gojala* is sprinkled over it.

Gojala is mixed with the seed. It is allowed to soak for 20-25 minutes and then dried. This is done every morning and evening for three days. Fresh *Gojala* must be used. The treated seed must be sown within three to four days. Too much *Gojala* should not be added because that may lead to germination of the seed before sowing.

Seed is winnowed. Light seed is discarded. Heavy seed is thought to be healthy and is used for sowing.

Cost

Only labour cost is involved. Gojala is available at the homestead.

Effect

Termites do not attack crop grown with seed treated with Gojala. The plant does not become yellow. It is healthier and gives better yield.

Festival

The treated seed is sown on auspicious days like Monday and Friday.

ANALYSIS

The above report is based on verbal descriptions given by farmers at Kalron, Lava and Phalodi. The actual use of this technique appears to be few and far between.

There is a growing tendency to buy packed seeds from the market, which is presumed to have been similarly treated.

SUGGESTIONS

There is a need to make a scientific study of seed treatment—particularly using Gojala. It has been reported that this technique has been experimented in some Gandhian ashrams. The results of these experiments have to be collected and analyzed. A pamphlet outlining the benefits can be made and awareness generated among the farmers.

Preparing *gojala*.

Gojala being collected.

Seed being treated with *gojala*.

3

Land Preparation (Sood, Jhoor and Kanabandi)

Category: Potentially Viable

SOOD

The field is cleared of natural plants and weeds like *Bui*, *Arana*, *Kheep* and *Ber.* The biomass is used for either *Kanabandi* or *Jhoor*. Removing the unwanted plants makes ploughing easy. The weeds compete with the crop for moisture and other nutrients in the soil. The weeds absorb nutrients meant for the crop if they are not removed. As a result crop production is up to 50 per cent less.

When is it used?

Sood is done about two or three months before the rains, i.e., in the months of May, June, and July. It is done when the farmer has spare time. In these months it is hot and dry hence there is little chance of the weeds growing back again after being removed once. Hence farmers do this work at their convenience.

Where is it done?

Sood is done in the fields where crop is to be grown.

How is it done?

Traditional tools like *Gandasi, Kassi, Jeyi, Kawadi, Bhola, Chousagi* and *Gada* are used to cut the weeds. All bushes and unwanted plants are cut. All members of the family, men, women and children, help in the work. Even old people join the work each working according to one's capacity. Sometimes even labour is engaged.

In level fields *Sood* is done before sowing so that fertile soil comes to the top. In uneven fields with *Tibaas Sood* is done in summers so that winds can make the field level. Such beneficial soil erosion exposes the roots of the plants and makes it possible for them to be removed.

In fields less uneven, small wild plants are uprooted after sowing so that the newly growing plants are protected from winds.

Cost

The only equipment involved is an ordinary farmer's tool. The tools are a *Kassi* costing Rs 100, *Gandasi* Rs. 50, *Bhola* Rs. 200, *Jeyi* Rs. 125 and *Chousagi* Rs. 125.

One labour is able to undertake *Sood* on one *bigha* land. The cost is Rs. 60.

The crop production is reported to be up to 30 per cent greater in fields where *Sood* has been undertaken. As a result, virtually all the farmers undertake *Sood.* An additional benefit is that the farmer gets biomass for undertaking *Kanabandi* or *Jhoor* or for use in constructing *Jhopa* for livestock or as domestic fuel.

Effect

The main benefit is higher crop yield. The secondary benefit is that uneven fields are leveled. Some farmers have reported that *Sood* helps in reducing the number of insects, which cause rashes and other diseases.

Women use the biomass as domestic fuel.

Festival

On the day of Akha Teej the tools are readied at an auspicious time and Sood is started thereafter. Sood is not done on days of *Amavasya, Shukla Ekadashi,* and *Poornima* as well as on Diwali and other festival days as farmers believe that harming insects in the soil on these days is inauspicious.

ANALYSIS

The beneficial effect of *Sood* is universally recognized. About 80 per cent farmers undertake *Sood.* The remaining do not undertake it because they are busy in wage labour or other works and are unable to supervise *Sood* work as well. This practice is likely to continue.

The modern alternative is to have the field ploughed by tractors using *tawi.* This deep ploughing upturns the soil and buries the plants. The biomass is still mixed in the soil and it does become manure howsoever unevenly. The difficulty is that *tawi* is best done every third year. Undertaking tawi every year is expensive. It also results in excessive loosening of the soil and consequently loss by soil erosion. Thus *Sood* remains an essential operation at least two out of three years when *tawi* is not applied.

KANABANDI

Kanabandi is a technique to trap the small organic particles that fly with the harsh summer winds. These particles act as manure and improve soil productivity. The technique helps present the loss of organic matter in one's own field through wind. It also helps in increasing the retention of rainwater though that is not the main purpose of the technique.

When is it used?

Kanabandi is done two to three months before the rains.

Where is it done?

Kanabandi is done in the field. It is done on the land where crop is grown and not on grazing land. For example if a farmer has 50 Bighas he may use this technique on 30 *bighas* where agriculture is done and not on the remaining 20 *bighas*, which are left fallow for grazing.

How is it done?

Branches of Khejri, Babool, Kair, Gwaar, and other trees are first cut and stacked at one's convenience. This is called putting up a *pai*. Then these branches are placed in a row on the borders of the fields before the rains thus forming a shallow protective wall about six inches to one foot in height. This is done in the north-south borders as usually the flow of wind is in the east-west direction.

Small particles of organic matter are trapped near the border due to obstruction as a result of *Kanabandi*.

Cost

The cost involved in *Kanabandi* concerns hiring of manual labour and purchase of a few ordinary agricultural implements such as a sickle and a large knife. To apply this technique to one *bigha* of land one has to spend Rs. 300 on the branches and labour.

It is very difficult to assess the increase in the yield due to *Kanabandi*. But it is clear that, unlike *Jhoor*, this technique has no negative impact. Hence the decision to undertake Kanabandi depends upon the availability of labour, or the opportunity cost of labour. Farmers rarely undertake *Kanabandi* where they have alternative wage employment. Often the old family members will do this work.

Effect

Kanabandi has three beneficial effects.

- The small particles in the wind are trapped as manure and that leads to higher productivity.
- The velocity of wind in one's own field is reduced and that reduces wind-based erosion of the topsoil.
- The *Kanabandi* slowly forms a sand border, which then prevents rainwater from flowing away.

ANALYSIS

Branches of trees like Babool, Kair and Ber are no longer available in plenty. The common grazing lands are not protected because of various reasons mentioned elsewhere. The available biomass is often used as fuel wood. This is the main reason for its decline.

Another reason for its decline is that the returns from *Kanabandi* are obtained over the next two or three years as the biomass gradually decays and becomes manure. *Kanabandi* is a type of investment. The returns from the immediate crop may not be substantial and people may not have the capacity to wait for returns that may accrue in future.

The third, though less important, reason is that people prefer to work in quarries in Jodhpur. The fact that wage labour is not engaged for *Kanabandi* indicates that the increase in crop production due to the use of this technology may be nominal.

According to Arjun Singh Rajput of village Digaadi, District Jodhpur, the practice of *Kanabandi* has dropped in the past 12 years because farmers tend to leave the village in order to work elsewhere and are not able to return on time to undertake *Kanabandi*. According to him, now only about 20 farmers out of 100 use the technique.

Kanabandi is being abandoned because there is the dual pressure of non-availability of biomass as well as of labour. It can be revived partly if alternate source of domestic fuel such as LPG or solar energy is made available. The problem of labour is more difficult. The returns from *Kanabandi* are low in comparison to the cost of labour. It is seen that farmers will do *Kanabandi* when they have spare time and no other employment. The availability of wage employment implies better incomes and it has largely contributed to the decline of *Kanabandi*. It is possible that there may be loss of soil fertility due to decline of this technique in the long run.

Kanabandi can be revived only if the returns are increased. This is possible only if the price of coarse food grains in general increase. It may also be possible to study and modify the technique to improve its impact on crop yield.

JHOOR

The biomass of small wild shrubs and grasses are cut into very small pieces and spread in the field before cultivation. Usually this is done with old biomass such as that which has been used in Baad and Jhopa.These pieces get covered in the soil during ploughing. Termites and other insects decompose them into manure leading to higher crop yield.

When is it used?

It is done before rains in the months of April and May.

Where is it done?

It is done in the field. The biomass collected from forests or grazing lands or from *Sood* is initially stacked in *Pai*.

How is it done?

As and when labour is available, the biomass of wild shrubs and grasses is thrashed or cut into small pieces, which are spread on the field. A long heavy thick wood is sometimes used as pestle to beat the biomass into small pieces.

Cost

Three days of labour is required to undertake *Jhoor* on one *bigha* land. The cost per *bigha* is Rs. 180.

The income depends upon the rains. If rains are less and much biomass has been spread in a small area of the fields then *Jhoor* becomes harmful. Too much of heat is produced in the soil due to decomposition of the organic matter. Therefore, *Jhoor* with rains is beneficial while intensive Jhoor with less rains can be harmful.

Effect

The land becomes more fertile as organic matter is added to it. If *Jhoor* is not done then the soil loses its fertility in six to seven years. The rainwater is better absorbed by the soil in fields where *Jhoor* has been done. Livestock can eat the leftovers of the biomass.

Festival

Sood is not done on days of Akshaya Triteeya, Amavasya, Shukla Ekadashi, Poornima, Diwali and other festival days, as farmers believe harming insects which live in the soil on these days is inauspicious.

ANALYSIS

About 15 per cent of the farmers undertake *Jhoor* at present. The rest are away working in the quarries of Jodhpur for extra income. They do not get the time for undertaking *Jhoor*. The technique is less frequently used because many families have started living in the villages rather than on the fields, which are far from home. It is not possible for them to use the spare time for *Jhoor*. Moreover, this work has to be done in the open fields in the summer months of April-June when it is very hot. The availability of biomass is another limitation as mentioned in the case of *Kanabandi*.

Jhoor is a labour-intensive activity. Trying to cut the biomass with the thresher or similar machines can possibly reduce the cost of this activity.

MODERN TECHNIQUE: CHEMICAL FERTILIZERS

The modern alternative to *Kanabandi* and *Jhoor* is to apply chemical fertilizers such as Urea and DAP. The difficulty, as in the case of *Jhoor*, is that chemical fertilizers have a heavy requirement of water. In absence of rains the fertilizers can burn the crop. Hence very few, if any, farmers are applying chemical fertilizers in dryland farming.

The cost of chemical fertilizers is less. Undertaking *Jhoor* in one *bigha* land requires labour input of three man-days or about Rs 180. A similar fertilization can be obtained by applying chemical fertilizers worth about Rs 25. The difficulty, however, is that chemical fertilizers are highly inflexible regarding the water requirement. Rains are a must within two to three days of application of chemical fertilizers. However, rains anytime within one month of sowing would be sufficient for *Jhoor*. Since rains are highly uncertain, there is not any practical alternative to *Jhoor* and *Kanabandi*.

SUGGESTIONS

One major problem is the availability of the biomass. Previously farmers would bring biomass from grazing lands. Now the grazing lands have mostly been cultivated and this source has dried up. Biomass is only available from wild plants and shrubs that grow in the fields. This biomass is very limited in quantity. Then there are demands on the biomass for use as domestic fuel and for making *bada* for livestock. These leave less biomass for *Kanabandi* and *Sood.* Possible action in this regard would be to encourage the use of alternate domestic fuel such as solar energy or LPG gas.

The second problem is that of availability of labour. The operations of *Sood, Kanabandi* and *Jhoor* are highly labour-intensive. For farmers the availability of work in urban areas is a more attractive and paying option than undertaking these activities. This problem can only be addressed by increasing the productivity of these activities.

1. Encourage tree plantation on field borders. This will have the same effect as *Kanabandi.* The main hurdle is giving water to the small saplings after transplantation. Experiments have to be done and costing worked out to see if the cost of giving water can be recovered. Various species of trees have to be tried in this experiment.

2. *Jhoor* can be done with *kadab*—the stalk of Bajra and other crops. Paddy straw is being regularly ploughed into the fields in Punjab. The leftover biomass of certain crops like Moong and Moth is not used as fodder anyway. A comparative study must be done to ascertain the returns from the biomass by its use as fodder and as *Jhoor*.

3. The biomass could be chopped in a thresher instead of manually. This may make it possible to the biomass chop into smaller pieces, which would increase productivity by allowing a more even spread.

4. Farmers have to be made aware that *Jhoor* should be undertaken over a wider area so that the danger from excessive soil heat due to low rains is minimized.

5. Implements for these operations may be modernized to increase productivity.

Land before *sood* is done

Sood in progress. Small wild plants are cut and removed.

Land after *sood* is done and mingani manure is applied

Biomass stacked for *jhoor*

jhoor in progress

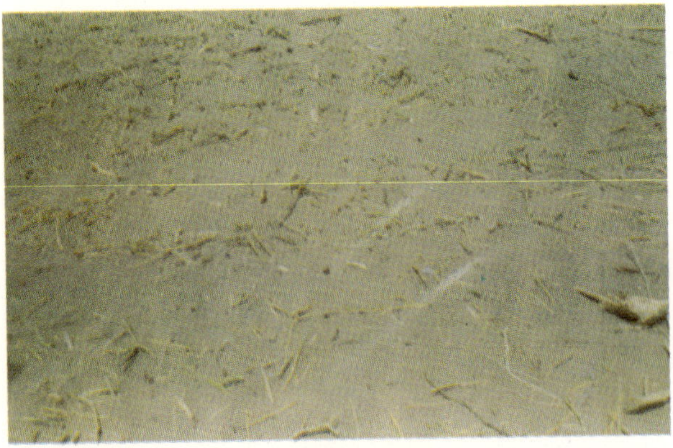

Field after *jhoor* is done

Excess *jhoor* leads to writing (middle rows)

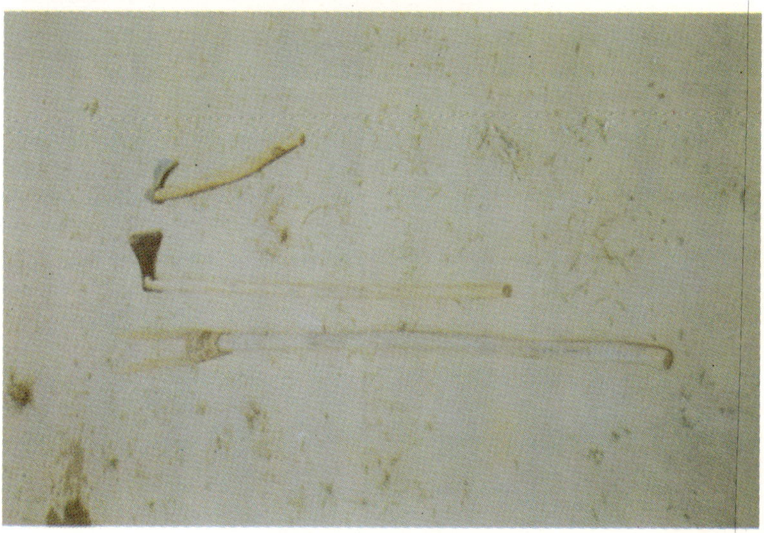

Implements for undertaking *jhoor*

Kanabandi has been done

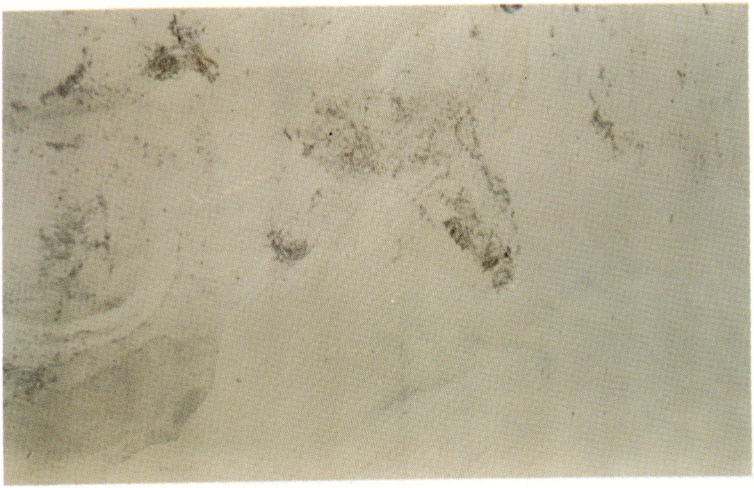

Small organic particles trapped by *kanabandi*

Closeup of organic matter trapped by *kanabandi*

4
Ploughing

Category : Declining

Ploughing is traditionally undertaken with camels and bullocks and, occasionally, donkeys. These livestock are integrated into the rural household in various ways. They provide milk and transport. The women fetch water and fodder for them. Lately tractors have substantially replaced animal power. In this section we examine the impact of this change on weeds, trees, crop yield and other aspects of crop agriculture.

When is it done?

The first ploughing is done in the month of *Ashad* or *Savan* (June-July) after the first rains.

Where is it done?

Ploughing is done in the field.

How is it done?

Ploughing is done with thehelp of camels, bullocks or donkeys. The farmer walks behind the plough holding it and also applying downward pressure on it so that the blade goes deep and makes the soil loose. With one hand the farmer holds the plough and with the other he drops seeds into the furrows created by the plough. Sometimes another person does the job of dropping seeds into the furrows. The seeds are sown in the east to west direction so that the crop allows the east-west wind to pass easily.

The field is first ploughed in the middle and then on the sides. The camel walks the length of the field and then takes a U-turn when it reaches the end. As a result, the edges of the field are unevenly ploughed and they have to be ploughed separately. Care is taken not to destroy plants growing on the borders of the field, which protect the field from winds and animals.

Ploughs of various designs are used with the draught animals (please see figs.).

Cost

A hectare of land can be ploughed in one day with a pair of camels or bullock. The cost of hiring a pair of bullock or a camel is Rs. 170 per day.

Festival

- Members of the family worship bullocks and camels before the commencement of ploughing.
- Bullocks are worshipped on festival days.
- *Khees* is prepared on *Akha Teej* and first offered to the bullocks and only thereafter partaken by family members.
- Small children undertake ploughing with toy ploughs on *Akha Teej*.

ANALYSIS

Use of animal power is threatened by the use of tractors. The cost of ploughing by tractor comes to the same Rs. 170 per hectare though the tractor undertakes this work in about an hour against the full day is labour put in by animals. The typical practice is to make deep ploughing with *tawi* once every three years and shallow ploughing with the cultivator in the intervening two years. Most farmers undertaking cultivation with bullocks use the services of the tractor every third year for *tawi* because it overturns soil up to a depth of about 12 inches against a depth of six inches by the draught animals. The effective choice, therefore, is limited to the use of tractor-and-cultivator or draught animals in the intervening two years. It seems that tractors have become a permanent fixture in the agricultural scene.

Quality of Ploughing and Crop Yields

The fertility of soil increases due to deep ploughing by tawi every third year. Farmers say that continuous use of *tawi* is also not beneficial as it loosens the soil too much leading to severe wind erosion. Kana Ram of Mandala estimated the average yearly yield of Bajra for different combinations of ploughing as follows:

Tractor : One year *tawi* plus two years cultivator

(20, 20 and 15 bags respectively in the three years) 18 bags

Tractor: Continuous cultivator 11 bags

Camel ploughing: Continuous camel ploughing

(yield is estimated at 18 bags less 2 bags due to delay in timely cultivation) 16 bags

Mixed : One year tractor *tawi* and two years camel ploughing 18 bags

The *tawi* holds the key. Deep ploughing once in three years overturns the soil and increases the absorption of rain water.

Depth of Ploughing and Weeds

The use of deep ploughing by *tawi* leads to less growth of weeds because the roots of weeds are destroyed. These weeds are used as fodder for the livestock. Their yield is about one-half after *tawi* is used. But this is a double-edged sword for it also reduces the workload of weeding, which is mostly undertaken by women.

The less growth of weeds is said to be one reason for higher yield from the use of *tawi* every third year.

The weeds are removed in *Sood* and used for *Kanabandi* or *Jhoor*. Less weeds, therefore, mean less biomass for these operations. On the other hand, *tawi* has the advantage of overturning and burying the entire weed that grows. The growth of weeds increases gradually in the two intervening years when *tawi* is not used . These weeds are entirely buried—a kind of mechanical *Jhoor* takes place during *tawi* operation. It saves considerable labour which is needed to undertake *Sood* and *Jhoor*.

Timely Sowing

A camel ploughs five *bigha* per day against five *bigha* per hour by the tractor. It is possible to sow large tracts of land with the help of tractors. Traditionally, the limiting factor for expanding the area of cultivation was the availability of draught animals. This limitation has since been broken.

Dry farming is critically dependent upon timely sowing. The seed has to be sown within four to five days of the rainfall. It is possible to cultivate only a limited area with draught animals in this limited period. As a result sowing is often delayed. Tractors are of much help in sowing on time and hence their use leads to an increase in yields.

Dung, Organic Manure and Fuel Cakes

The dung of bullocks and camels can be used as manure. But farmers have to depend more on chemical fertilizers if they use tractors.Chemical fertilizers however, require timely irrigation hence they cannot be used in dry farming. The result is that dung and manure from draught animals is no longer available. However, the fodder that is grown as stalks of Bajra is fed to milch animals—buffaloes and cows, which provide the same dung. The limitation in the availability of dung manure is the availability of fodder, not the number of draught animals. Hence it appears that there is no detrimental effect on soil fertility on this count.

Saving Trees and Shrubs

Tawi cultivation ploughs land up to twelve inches deep and it cuts up rootstock of Khejri and other trees and shrubs, which may be growing. Secondly, it is not possible for the tractor operator to see and avoid small single stalks of shrubs or trees. The farmer is easily able to save only individual trees and shrubs. But it was found that even where the tractor driver could veer around and save a tree, he would not do so. The farmers explicitly said that they did not want trees in their fields, which had an impact on the valuable yield of food grains.

Why should they have tolerated the trees earlier then? It seems that the traditional limitations was that of draught animals. The number of draught animals determined the area that was cultivated. Since land was not the limitation, the farmer had nothing to lose by saving the trees. Cultivation by tractors has removed the limitation of draught power. Now land has become the limiting factor. Hence farmers are not inclined to save the trees. We can conclude that the loss of trees and shrubs is not fundamentally occurring due to cultivation by tractor, rather it is occurring because the farmer does not like the trees.

But the Khejri tree is an exception to this rule. There is a general consensus among the farmers that the number of trees—especially those of Khejri—growing in the fields is becoming less with time due to the use of *tawi.* The shrubs—Ber, Sinia and others—do grow gradually in the intervening years but there appears to be a large decline in the population of Khejri trees. Khejri is agro-friendly. It does not compete with Bajra for land. It provides excellent fodder and fuel wood during annual lopping and timber when it is finally cut. The loss of Khejri is a major factor that merits consideration.

Evenness of ploughing and sowing

Sowing is done by tractors with the farmer sitting behind on top of the cultivator and dropping seed manually. This leads to uneven seeding. Patches of land are left unseeded while other parts are overseeded.

Effect on Women

Women bear much of the responsibility of taking care of the livestock—

chopping fodder, fetching water, giving them a bath and even taking them for grazing. Tractors reduce the workload.

Women often drop seed when draught animals do sowing. Sowing by tractors reduces this workload of theirs.

It is not necessary to clean the field of grasses before ploughing (*Sood*) and also to chop the shrubs and spread them for use as fertilizer (*Jhoor*) if cultivation is done by tractors. This reduces the workload.

Others

The shift from bullocks may also lead to removal of cows and buffaloes from the homestead. If that happens then the household is deprived of milk, dung for fuel as well as smearing on the walls and courtyard, and gojala and camel urine, which are used for treatment of diseases. But it is not necessary to remove the milch animals along with draught animals. In fact, it is seen that farmers are maintaining milch animals in such numbers that can be sustained by their own availability of fodder. There is a need, however, to remain vigilant on this issue.

The draught animals can also be used for travelling and transporting goods to and from the markets and for bringing water for household purposes in the summers. Tractors are used profusely in the Bhadariya Oran. On being asked, Bhadariya Maharaj said, "The bullocks are made to work very hard. They should not be so subjected to work".

Conclusion

It seems that the use of draught animals for ploughing will gradually come to an end. The main factor for this appears to be that of timely operation. The available land area cannot be cultivated in the short time that is available and with the number of draught animals that can be put to use. The basic problem is fodder from a hectare of land cannot sustain draught animals required for its cultivation. The fodder requires to be supplemented. The farmers could undertake cultivation with draught animals as long as common grazing land was plenty and such supplement was available. The situation has changed now and such supplement is not available. Hence, cultivation from draught animals is not likely to sustain. Improvement of breeds, it appears, can help only marginally. Not a single farmer was found to undertake all the cultivation with draught animals.

The availability of wage employment in urban areas further reduces the incentive for keeping draught animals. Men prefer to work in the cities while women take care of the fields. Men like to come in the sowing season, get the cultivation done by tractors and return quickly.

The combination of one-year of the *tawi* and two-years of the cultivator appears to be suitable in the present conditions. The situation may change substantially in the very long run when the supply of fossil fuels is exhausted. But then mankind may come up with newer solutions at that time.

SUGGESTIONS

The following measures may be attempted to reverse the decreasing use of draught animals.

- Breed improvement. Try to find or develop a breed of cows or buffaloes which double up as draught animals for short periods.
- Improved ploughs. Increase the efficiency of draught animals.
- Crop selection. Increase the amount of fodder available from Bajra and other crops. This will make it possible to maintain a higher number of livestock.

Ploughing by camel

Ploughing by single bullock

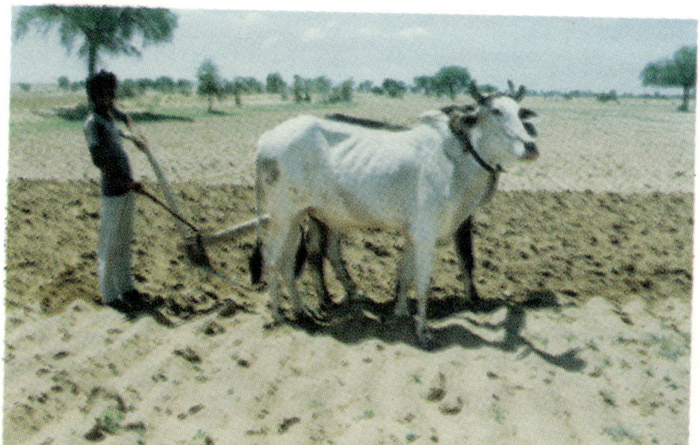

Ploughing by two bullocks

Ploughing by donkeys

Children undertaking ritual ploughing with a toy plough on Akha Teej

Uneven sowing by tractor

This shrub could easily be saved during ploughing be tractor

Tractor driver demonnstrating how shrub can be saved if desired

The *tawi* (plough) in operation

Farmer pointing out the maximum depth of a camel plough. The tractor has just ploughed with *tawi*

Grass grows in field where *tawi* has not been done

No grass grows where *tawi* has been done

Camel plough

Camel plough with two blades

Camel plough with two blades

5

Re-Sowing (Bhanj)

Category: Stable

The seed may sometimes not germinate evenly due to poor rainfall or other reasons. The farmers undertake sowing once again in such a situation. They try to save as many of the previous germination while sowing more seed in - between. This operation is called *Bhanj.* The idea is that the few plants that have geminated should be allowed to grow while the ungerminated areas alone are re-sown.

When is it done?

Bhanj is undertaken about three to five weeks after the first sowing.

Where is it done?

Bhanj is done in the field.

How is it done?

The farmer takes his plough over the middle of the two rows sown earlier and seeds the field once again.

Cost

The cost of a bullock or camel is about Rs 100 per day. About 2 *bigha* of land can be subjected to *Bhanj* in a day. The cost of *Bhanj* is about Rs. 50 per *bigha.*

Effect

More plants grow in areas which were hitherto empty. This leads to higher yields.

ANALYSIS

Draught animals best undertake Bhanj. The single blade of the plough can easily be steered to save plants that have germinated. It is not possible to save old plants in tractor cultivation. However, farmers are undertaking *Bhanj* with tractors as well. Large patches in a field where seeds have not germinated are re-sown (please see figure).

SUGGESTIONS

The traditional system of *Bhanj* will continue as long as draught animals are available.

It may be possible to devise new types of cultivators for use with tractors in which an operator sitting on the cultivator may have the ability to change the position of the blade to save existing plants. Framers have already started changing the number of spikes in the cultivator. Special spikes could be designed for *Bhanj*.

Bhanj in progress by bullocks

Bhanj in progress by camels

Bhanj done by tractor

6

Mixed Cropping

Category : Stable

Mixed cropping or *mishrit kheti* means growing two or more crops simultaneously in the same field. Crops like Bajra, Guwar, Jwar, Til, Moong, Moth, Chola and Matira are grown together.

When is it done?

Seeds are sown at the onset of rains.

Where is it done?

The farmer sows the seeds in his field.

How is it done?

Seeds of different types—up to three or four types—are mixed together and sown. Different crops grow in the same field together.

Effect

The farmers mostly depend on rain for their cultivation. Mixed cropping is a kind of insurance. If rains are more, one crop will give good results. If rains are less, another crop will give good results. In either case some produce is assured.

Fertility of the soil increases by mixed cropping. For example the leaves of Moong, Moth and Guwar fall and make the soil fertile.

Cattle harm the crop less in mixed farming. If cattle enter the field it might prefer to eat a particular crop only. But as it would not find all the plants eatable, less harm would ensue.

The impact of mixed cropping on crop diseases is mixed. In some cases the disease of one crop spreads to another. such as the disease of

Gondia that is common in Til spreads to Bajra. Similarly the pest *Katra* is first attracted by Moth but then it attacks Bajra as well. In other cases the diseases of one crop do not spread to another and the latter is saved.

Different kinds of feed are obtained. Mixed together, they emerge more nutritious for the cattle.

The erosion of soil is less. The root pattern of different crops compensates for the peculiarities that one or more crops may have.

A distinct disadvantage of mixed cropping is that more hard work is involved especially at the time of harvesting. Different plants have to be bypassed while harvesting others.

Cost

Many combinations of mixed cropping are possible. One farmer, for example, mixed the seed of five crops for sowing as follows:

Bajra	2 kg @ Rs. 8 per kg	Rs. 16
Guwar	2 kg @ Rs. 10 per kg	Rs. 20
Moong	1 kg @ Rs. 14 per kg	Rs. 28
Til	½ kg @ Rs. 26 per kg	Rs. 13
Matira/Kachara	¼ kg @ Rs. 32 per kg	Rs. 8
Total		**Rs. 85**

Mono Cropping (Irrigated)

Farmers universally grow single crop on irrigated lands. The reason for adopting mixed cropping is to insure against variation in rainfall. When irrigation is assured the very cause of undertaking mixed cropping is eliminated. In fact, mixed cropping in irrigated fields would be harmful. One crop would be irrigated according to the requirements of another. A survey found that *all* farmers were practicing single cropping on irrigated fields.

One example of 'mixed' irrigated cropping is observed in chilies in the middle of which castor is planted to protect the chili crop from *Jholaa* and *pala* (hot winds and frost). However, this is not really 'mixed cropping'. Castor is not sown for producing castor seed. Rather it is sown to protect the chili crop.

ANALYSIS

Mixed cropping is the norm for rainfed agriculture. A survey found that 75 per cent of farmers were using this technology.

Survey

Technique name	*Number of respondents using*	*Reasons*
Mono Cropping (Irrigated)	10	The crop gets water and fertilizer according to its individual needs.
Mixed Cropping (Irrigated)	0	Some crops need more and some less of water and fertilizer.

Survey

Technique name	*Number of respondents using*	*Reasons*
Traditional: Mixed cropping rain Fed	6	Many crops can be had from one field. Livestock have fewer tendencies to graze such crops hence the cost of protection is less.
Modern: Mono cropping	2	Harvesting the crop requires more labour.

Mixed cropping is mostly undertaken in rainfed agriculture—mainly as insurance against erratic rainfall. The farmers invariably shift to monocropping in irrigated agriculture. If the farmer can provide irrigation then mixed cropping would require irrigation as per the requirement of different crops. Bajra would be given irrigation merely because the Moong that has been cropped needs it though Bajra itself does not need the same.

Other advantages of mixed cropping are as follows:

- Crops growing together protect each other. Tall crops are naturally able to withstand the winds and they protect smaller crops.
- If some animal enters the field then too the crops could be protected. The animal might prefer to eat a particular crop, say Moong. But as it would not find the Moong plants together there would not be much harm done.
- If some disease attacks one crop and another crop survives and grows well the farmer does not have to suffer total loss.

SUGGESTIONS

The farmers are regularly making innovations and experimenting with different crop mixtures. The only hurdle for promotion is the cost of labour during harvesting. The objective of agriculture is to provide the farmer with a decent income. There is, then, a direct conflict between traditional technology and income.

Theoretically, it should be possible to undertake experiments with different mixtures. But this is not practical because the variation in soil and rainfall conditions is very large. Any experiment would, therefore, have limited applicability.

However, it may be useful to document the benefit of mixed cropping and submit a report to the agricultural universities or organize a seminar to encourage them to undertake plant breeding keeping in mind the synergy of different crops for mixed cropping.

Mixed dryland cropping

Tall plants of til amidst moong, moth and guwar

Mixed dryland cropping. Small plant of til (in front of the lady) amidst field of bajra

Mono cropping of cotton on irrigated fields

Mono cropping of bajra (on left) where assured irrigation is available

Castor plants amidst chili fields. This is not irrigated mixed cropping. Castor is planted to protect from frost, not as a crop

Mono cropping of chilies on irrigated fields. Castor on field borders for protection from cold winds.

7

Cultivation in Bund Bed (Khadin)

Category : Potentially Viable

In rainy season water gets collected in *Khadin* and it is slowly absorbed by the soil. Bajra is sown in the *Khadin.* If rain is less then Bajra grows well. If rain is heavy then Bajra in the bottom of the *Khadin* may be lost due to flooding but the land absorbs water and a crop of mustard or wheat can be taken in *rabi*. The *rabi* crops are generally irrigated. But they grow well in a *Khadin* without irrigation because of the moisture absorbed by the land during the rains.

When is it done?

The main Rabi crop is sown in the *Khadin* in the month of October or November after Diwali.

Where is it done?

Crop is grown in the *Khadin*. It is done in the bed of the *Khadin* where the water has collected.

How is it done?

Normal crop cultivation is undertaken—nowadays by tractors.

Effect

The soil of *Khadin* is more fertile as the organic matter coming from the catchment area is held up here. There is no need to irrigate the crop.

Another benefit of *Khadin* is that it provides drinking water too. The water table rises to 15-20 feet near the *Khadin.*

Cost

Normal crop cultivation is undertaken. No additional cost is involved.

ANALYSIS

Those who have *Khadin* undertake cultivation in *Khadin*. There is no 'modern' alternative. The peculiar feature of this cultivation is to grow Bajra in the *kharif* season. If rains are less then Bajra grows nevertheless because it gets more water than the normal field. If rains are good then some Bajra may die because of flooding but the good rabi crop that follows more than compensates the loss.

Suggestions

Quick maturing varieties may be promoted in *Khadin*. Such varieties should be located that can mature before the moisture in the bed dries up.

Water resistant varieties of Bajra may be searched or bred for cultivation in the bed of *Khadin*.

Cultivation in khadin

VII

Crop Husbandry

8

Weeding (Nidaan)

Category: Stable

The removal of unwanted weeds amidst growing crops is known as *Nidaan*. Small weeds and grasses come up along with the crop. They compete with the crop for the limited available moisture. The weeds are uprooted in *Nidaan*.

When is it done?

Nidaan is done one to one-and-a-half months after sowing when the crop is about one foot high.

Where is it done?

Nidaan is done in the crop field.

How is it done?

A particular implement called *Kassi* is used to uproot the plants. *Kassi* is made of iron and has a wooden handle. The farmer stands, hits with the *Kassi* at the root of the weed, and pulls it up. The weed or grass is uprooted and left in the field. The weeds may be picked up and used as fodder for the livestock. Grasses, which are not usable as fodder, are left to dry in the field and abandoned. To collect such grasses would be an unnecessary work. The same grasses would be collected before ploughing, and cut and spread in the field under *Sood* and *Jhoor*. But after sowing, the non-edible grasses cannot be collected, cut and spread this way. They can only be stored till the next season. This is not done.

Effect

Nidaan ensures that moisture in the soil is used by the crop and not by the weeds. Farmers estimate that up to 25 per cent higher yields can be obtained by undertaking *Nidaan.* Edible grasses are used to feed the cattle. This is very important because other sources of fodder are not easily available at this time.

Cost

It takes about one manday per *bigha* to undertake *Nidaan.* Of course, it depends also on the intensity of the weeds. At a wage rate of Rs. 60 per manday the cost comes to Rs. 60 per *bigha.*

The yield of Bajra is around 100 kg per bigha. At 25 per cent increase in the yield, which is the maximum, the increased production may be about 25 kg valued at about Rs. 100. This is the best-case scenario. The costs and returns are evenly balanced. Therefore, farmers will generally undertake Nidaan only if wage employment is not available for them.

LAAH

Laah is a generic word used for collective work on various occasions such as weeding, harvesting or making a house. One farmer invites other farmers and neighbours to come and help in the task on a particular day. At least one person from every family comes to help because it is considered a moral commitment. The work is then reciprocated. The host will go to the field of the others on selected days and do the *Nidaan.* The benefit of collective work is that farmers are able to sequence the weeding to match the growth of weeds. The field in which weeds are more may be taken first and so on.

When is it done?

Laah is undertaken when farmers have large fields which are difficult to manage alone. *Laah* can go on for one or two days.

Where is it done?

The farmer does *Laah* in his field.

How is it done?

The farmer consults his family and decides on which day they should call upon the people of the village to help. Then the farmer announces in the village that *Laah* would take place in his fields on a particular date. On the designated day when the people reach the field first they are given tea and breakfast. Then they go to work. In the daytime they are given food to eat. Then again they work and in the evening after having food again they leave for their homes. The farmer in whose field the work is going on feeds the workers. If the farmer is rich more people turn up in the hope of getting better food than when the farmer is poor.

Effect

Laah in the field is done quickly and completed at one go. Undertaking this task individually may lead to weeds growing large in one part of the field while the farmer is busy in another part.

Women have to prepare food for 20-50 people. This can be a tiring job. But there is a benefit too. The women too would be undertaking the weeding in the field. They have to prepare food instead of undertaking *Nidaan.*

Laah helps build cordial relations among the villagers and fosters the spirit of community as they work together with cooperation. The good food served in Laah is like a feast and an occasion of joy.

Cost

Laah is cheaper than hiring labour as only the cost of the food is borne by the farmer. Most of the ingredients for the food are homegrown hence the cash expenditure is less. Previously only *khichadi* (porridge) was given to the people who came to work. But nowadays good food and *Halva* is served. Tea, tobacco and *bidis* are also offered. Perhaps this is an indicator of increasing level of opportunity wages that are foregone.

Festival

Laah is done with a spirit of festivity. People sing folk songs while working. One such song is as follows.

Tthandi Re Belaa Jor Am Bhaaiyo,
Tthandi Re Belaa Ro Kaam Bhaaiyo,
Dinado Re De Gayo Dor Ra Am Bhaaiyo,
Bahaate Re Kinee Moj Am Bhaaiyo,
Bheerudo Re Bhelo Beve Am Bhaaiyo,
Meettho Re Bolo Raam Am Bhaaiyo,
Gududi Re Birkhaa Jor Am Bhaaiyo,
Ghurane Re Bhariyo Der Am Bhaaiyo,
Dedar Jal Ro Jeev Am Bhaaiyo,
Levasi Re Jal Ro Tthaag Am Bhaaiyo,
Meh Butthei Re Huve Ler Am Bhaaiyo.

(The times are cool O friends,
It is time to work in this coolness O friends,
The sun is setting O friends,

Let us sit and rest O friends,
Let us all work together O friends,
Speak Ram in sweet vc ice O friends,
The rain is falling fast O friends,
It thundered and rained and filled the fields with water O friends,
Frogs are creatures of water O friends,
People feel joyous in rain O friends,
Youth always yearns to work O friends,
All repeat the name of Ram O friends,
It is work for many O friends,
The sun has set O friends.)

ADI BADI

Farmers cooperate with each other to cut the crop. There is a mutual arrangement to help each other. Farmer A works in the field of farmer B for a day and vice versa. This arrangement makes it possible to complete harvesting of one field quickly. It is also convenient to bring heavy loads of *kadab* which are carried together rather than singly.

In *Laah* there is no binding for one to reciprocate. The food and feast are considered to be adequate and final compensation. In Adi Badi there is a mutual contract to return an equal amount of labour. All other details remain the same as in *Laah*.

WEEDICIDES

Many farmers also use chemicals in the form of weedicides to destroy unwanted plants. This is mainly done for the *rabi* crop of mustard (Rayada). This crop is profitable. It is grown only by a few persons who have facilities for irrigation; hence Laah and Adi Badi are not possible in this case.

When is it done?

It is done generally in November when the crop has grown.

Where is it done?

It is done in fields where Rayada crop is growing and there are weeds.

How is it done?

A solution of weedicide is prepared as prescribed. Then it is sprayed over the crop.

Cost

The cost for one *bigha* field of Rayada is estimated as follows:

Cost of weedicide	Rs. 40
Cost of one manday labour	Rs. 60
Total	**Rs. 100**

Effect

The weedicide destroys the weeds and unwanted plants and the crop grows well. But the weeds are completely destroyed and there is no grass or plants left for the cattle to feed upon.

The chemicals have to be used carefully or they can be harmful for the user.

ANALYSIS

Not one farmer used weedicide in the *kharif* crop of Bajra. However, a survey shows that about one-third did the same for growing Rayada.

Survey

Technique name	*Number of respondents using*	*Reasons*
Traditional; *Nidaan*	16	The grass and weeds are used to feed cattle.
Modern; Weedicides	7	The weeds get totally destroyed and the crop grows well. It takes less time than *Nidaan*.

Kharif (Bajra)

The central question is of wage rate. The returns from *Nidaan* are marginally poised. If the crop yield and price is good then it is profitable. If the opportunity wage foregone is low then it is profitable. An increase in availability wage employment and wage rate will bring forth a reduction in the use of the technique and also *Laah*, which is but a supporting social system for *Nidaan*. Farmers may then simply leave the weeds and accept a lower yield. Otherwise the technique is stable and no modern substitute is visible.

Rabi (Rayada)

Nidaan is not usually undertaken in the *rabi* crop of Rayada for the following reasons:

1. The crop of Rayada is irrigated repeatedly and thus the weeds regrow fast. *Nidaan* has then to be done repeatedly.
2. There is growth of insects that renders the weeds unfit for use as fodder.

SUGGESTIONS

A comparative study of the long-term effects of weedicides and the traditional *Nidaan* on land and environment may be undertaken. If adverse effects are substantiated then Nidaan may be propagated among the farmers.

Nidaan in operation

Wild plants uprooted by *nidaan* and left to dry in the field

Lah for *nidaan*

Food preparation for *lah*

9

Protection from Frost by Watering (Pala)

Category: Stable

The crop is watered when there is fear of frost (*Pala*). This prevents the crop from being destroyed by Pala. Smoke is created in the field in the evening. This provides warmth to the plants and prevents damage from *Pala*.

When is it done?

It is done in the winters, January in particular, when it is very cold and there is fear of frost descending in the night.

Where is it done?

The technique is used in one's field where the crops have been grown.

How is it done?

Water is sprinkled in the field with the help of sprinklers. Or the field is irrigated. Smoke is created in the field in the evening.

Effect

The crop is protected from frost. But this technique leads to more use of water and more of underground water is exploited.

All family members help in the irrigation of the fields. Even women help in this task.

Cost

No cash cost other than the normal costs of irrigation is incurred.

ANALUSIS

This technology is used in winter crops that require irrigation such as mustard and chilies.

Survey

	Number of respondents	*Reasons*
Irrigating the crop for protecting it from *Pala.*	14	Irrigating the crop in winters and creating smoke saves the crop from *Pala.* It is a cheap technique that saves the crop.
Not irrigating the crop for protecting it from *Pala.*	Nil	

SUGGESTIONS

Studies can be made regarding the type and intensity of irrigation required for protection of crops from frost. The farmers do not make a distinction between normal irrigation and irrigation for 'frost protection' purposes.

Protection from frost by watering

Protection from frost by watering

10

Protection from Frost by Castor Plants

Category: Stable

Arandi or castor is also grown along with crop of chilies to protect the latter from frost (*Pala*). Castor plants are able to protect 60 to 70 per cent of the chili crop from damage. Castor plants are grown on the sides of saplings of chilies to protect them from hot winds called *loo* and the heat of the summer.

When is it done?

Seeds of Castor chilies are sown in the month of August in the fields where chilies are to be planted.

Where is it done?

Castor seeds are sown on the sides of the furrows where chili saplings are to be planted.

How is it done?

When furrows have been prepared for planting chili saplings then on the edge of each furrow the castor seeds are sown at a distance of three feet from each other. When the castor plants grow their leaves are cut so that they grow taller to protect the chili plants better. Sometimes only one row of castor plants is planted and that on the windward side of the field. This provides limited protection.

Effect

The tall plants of Castor act as a windbreak and prevent cold winds carrying frost from moving close to the chili plants. The crop of chilies is thus protected from *Pala*. Though the castor plants draw nutrition from the land and adversely affect the chili crop this loss is compensated by the protection they offer to the chili plants from the frost. The crop of chili is about 10-15 per cent less due to the growth of castor plants alongside.

The leaves of the castor plant are added to the fodder for livestock. Their wood is used for making fire and also to make *Jhopa* (huts) and animal shelters. castor seeds are sold.

Cost

About 1 g Castor seed costing Rs. 50 is required for growing chili in one *bigha* of land.

ANALYSIS

All farmers growing chilies also grow castor as per the survey.

Survey

	Number of respondents	*Reasons*
Farmer growing castor with chilies	14	The crop of chilies is protected from *Pala*. The leaves of castor plant are fed to cattle. One gets wood for burning from the castor plants. Seeds of castor are also obtained which are sold at a good profit. The branches and wood of castor are also used for making *Jhopa* and shelters.
Farmers not growing castor with chilies	Nil	The loss to crop of chili would be much greater without protection from castor.

SUGGESTIONS

A study of varieties of castor may be made from the standpoint of low consumption of water. The varieties available in the market have been developed to enhance yield of castor seed. These may not be the most appropriate varieties for use as protective walls.

The government must make modern facilities of communication available so that the farmers could receive information of forecasts about *Pala* made with the help of satellites. The village *Panchayat* and the mass media should take the initiative of informing and educating the farmers.

Castor plants in chili fields for protection from frost

11
Insecticide

Category: Potentially Viable

Insecticides are used to protect the crop from fungus, insects and other diseases.

Traditional Insecticides

When are they used ?

Some farmers make preventive use of insecticides. Others use insecticides after the attack of diseases.

Where are they used?

The farmer uses insecticides on the crops in his field.

How are they applied?

There are several traditional methods of using insecticides:

- Smoke is generated by burning bones of camels, droppings of goats and dry dung of cows if the crop of Til and Guwar is attacked by the insect *Teiliya*. This smoke kills the insect and the crop is saved.
- The sand from cowsheds is mixed with ash and sprinkled over the crop to protect it from insects.
- A trench is dug around the field so that the locust (*Tiddi*) and its young ones fall into it. Then the trapped locusts are covered with soil or ash and smoke is also made in the trench to kill them. Mostly locusts lay eggs in trees of Kair. Farmers put dry dung into the holes where these eggs are laid and set the dung on fire thus destroying the eggs.

- Fire is lit in the fields in the night to protect the crop of Bajra from Katara. The fly-like insect falls into the fire and gets burnt. The problem, however, is that if only one farmer makes the fire than the Katara from the surrounding fields also come into his field but not all get burnt. This technology is very effective when undertaken collectively.

- Tantra was also used earlier for ridding the crop of insects. In old times to protect the crop from an insect called *Karva* a Tantra ritual was used. The *Bhanja* (son of one's sister) would sit on the shoulder of the *Mama* (maternal uncle) and eat *Choorma* at the same time repeating the words given below. Javaram of village Kan Singh Sidd said that once he actually used this technique seated on the shoulders of his uncle Paburam.

  Mama Chad Bhanja Aya

  Utth Karva Khet Paraya.

  (Rise Mama, your Bhanja has come/ Rise and have the field cleared.)

- Ash is sprinkled over the crop in early morning when moisture from the dew is present. The ash sticks to the leaves and protects them from the insects.

- Two persons hold the ends of a long rope and they walk fast across the field. This drives away the locusts. This is done again and again.

- A mixture is prepared out of cow urine, Neem plants and Aak leaves. Cow's urine is collected. Leaves of Neem and Aak are soaked in the urine in a pot and covered with a lid. In 15 days the insecticide becomes ready. This is sprinkled on the crop to protect it from termites.

MODERN INSECTICIDES

Many farmers buy insecticides from the market and use them.

- The farmer buys insecticides available in the market and sprays the prescribed quantity over the crop with a spray machine. In case there is an 'epidemic' then the government makes the insecticides available to the farmer.

- The government sprays DDT and other insecticides over the crop by airplanes if locusts attack the crop.

Cost

The traditional technologies mainly use items available within the homestead such as Neem leaves and cow urine. The main cost is that of labour. Modern technologies, on the other hand, require cash purchases of insecticides. The cost of spraying is similar in both technologies.

Effect

- Land is used to dig a trench to trap locusts and Katara.
- Some insecticides are reported to increase the water requirement of the crop.
- Chemical insecticides available in the market are poisonous. They have to be used carefully. Cattle are not allowed to graze on the land where insecticides have been used.
- If airplanes sprinkle insecticides, water in water sources like ponds gets poisoned.

ANALYSIS

Survey

Kind of Technique	*Number of respondents using*	*Reasons*
Traditional; Home- made insecticides	4	They are cheap and involve no cash expenditure.
Modern; Chemical insecticides bought from more market	5	Chemical insecticides, are more effective. Do not know how to prepare traditional insecticides, hence use chemicals.

The main question is that of reliability. Farmers generally report that modern chemicals are more effective. The difference between the two is as follows:

Traditional : No cash expenditure, labour intensive, less effective.

Modern : Cash expenditure, less work, more effective.

It seems that the decisive factor is that of reliability. The crop is the lifeline of the farmer. He is generally willing to find the cash required if he is assured of protection.

SUGGESTIONS

- Comparative field trials of traditional and modern insecticides should be undertaken. The results should be published as a pamphlet and disseminated among the farmers. The studies should give special attention to the development of immunity among the insects, which is believed to be more pervasive to modern chemicals. Negative environmental impact such as denuding of neem trees for their leaves and pollution of water bodies should also be included in the study.
- The details of traditional technologies should be collected from universities and other institutions. The Centre for Indian Knowledge Systems, Chennai has a programme of studying *Vriksha Ayurveda.* Treatises have been written on plant treatment in the Indian tradition. They may be compiled and their field trials undertaken.
- A programme to create synthesis of traditional and modern treatments may be undertaken. Ayurveda is reported to be more effective for certain diseases among human beings while allopathy is reported to be more effective for others. Similar combinations can be tried for plant treatment.
- Effort must be made to provide certified ingredients for the use of traditional technologies.

Katara (orange colour, in middle) crossing a baad

Traditional insecticide gojala, neem and aakh being prepared

Ash sprinkled on garlic crop

Modern chemical insecticide being sprayed on cotton crop

12

Insecticide (Ash in Jira)

Category: Potentially Viable

A mixture of ash and sulphur is sprinkled upon the crop of cumin seed (jira) to prevent it from getting affected by the disease of white spots (Gundiya and Phankudi) and by an insect called Al.

When is it used?

Sprinkling is done when flowers start to appear in the Jira crop.

Where is it done?

It is done over the crop of cumin seeds.

How is it done?

Ash and sulphur, which has been finely powdered, are mixed together and then the mixture is sprinkled over the crop by hand in the early morning. The mixture is put in a sack of fine cloth and the sack is carried through the field allowing the mixture to fall over the crop when the sack is beaten. The ratio of sulphur to ash varies between 10 and 50 per cent depending upon the intensity of the disease. More sulphur is used for greater effect.

In some places the farmers repeat the treatment every 10 days. In some villages like Gagadi the technique is used in one cycle while in other villages like Kalron it is done four times. May be this is due to the difference in the type of land.

Effect

The crop of Jira is protected from diseases and insects. There is about 40 per cent more produce if the technique is used.

Sulphur and ash do not have any negative effect on the soil. Rather the ash makes the soil more fertile.

Caution is needed in the use of sulphur. The crop must be irrigated on the third day after sprinkling otherwise the plants dry up and die. If the cattle eat the plants on which ash and sulphur has been sprinkled then they can get ill and even die.

For the family the benefit is that the technique is easy and the ingredients, i.e., ash and sulphur are easily available. Women mostly do the task of sprinkling the mixture so their work increases.

Cost

For a field of one *bigha* 2 kg sulphur is needed each time for sprinkling to be done four times.

Cost of 8 kg sulphur	@ Rs. 10 per kg	Rs. 80

Festival

There are many folk songs related to the crop of Jira, which are sung on all festive occasions.

One song is

'Mat Bo O Jira

(Do not sow the crop of Jira)

This song hints at the various diseases that could attack the crop of Jira and it cautions the farmer against sowing this crop in his field.

Chemical Weedicides

At present instead of sprinkling a mixture of ash and sulphur farmers spray a chemical called **Finadol**. This chemical protects the crop from disease and the produce is more.

When is it used?

It is used when the insect Al or some disease afflicts the crop. It is generally done 90 days after sowing the seed.

Where is it done?

It is done on the crop of Jira.

How is it done?

The chemical Finadol is mixed in water and it is sprayed over the crop with a spray machine. Or alternatively the powdered chemical is put in a sack or cloth with fine holes and then the chemical is sprinkled over the crop by beating the sack. The crop should not be watered for a week after the use of this technique.

Effect

A solution of the chemical has to be prepared in water. Hence care must be taken, as the chemical is poisonous. Men do this job so women are exempted from this work.

Cost

In 1 *bigha* field about 4 kg Finadol is needed for one sprinkling. It may be repeated two three times.

Cost of 4 kg @ Rs. 20 per kg for 2.5 times Rs. 200

ANALYSIS

The cost of Finadol is about double that of ash and sulphur. A survey shows that about one-fourth of farmers are using Finadol instead of ash and sulphur.

Survey

Name of Technique	*Number of respondents using*	*Reasons*
Traditional Ash and sulphur	17	Use of ash and sulphur is cheaper.
Modern: Finadol	4	The insect Al does not attack the crop. If afflicted, the crop becomes disease free with the use of the chemical Finadol.

The decision to use Finadol or ash and sulphur depends upon the intensity of the disease and the ability of the farmer to bear the higher cash expenditure of purchasing Finadol.

SUGGESTIONS

The ratio of ash and sulphur, the intensity of sprinkling and the time between two sprinklings may be scientifically studied and the learning may be shared with the farmers.

A comparative study of the two methods may be undertaken. The results may be propagated for the benefit of the farmers.

Ash and sulphur being sprayed on jira crop

13

Protection from Birds

Category: Potentially Viable

Birds are a nuisance for the farmers. They eat the ripe grain and destroy the crop. The crop of mustard (Rayada) has to be specially protected from birds. When the seed of Rayada is sown it is irrigated. After three days the seeds germinate and saplings spring out. This is the time when birds have to be driven away. The birds uproot the young saplings and destroy the new crop.

When is protection from birds needed?

Birds are driven away when seeds have been sown, when young saplings have appeared and when the grain in the crop has ripened.

Where is it required?

It is required in the field.

How is it done?

Farmers use many methods to drive away the birds.

Catapult (Gafni)

A piece of stone is put in a leather holder, which is swung around. One string of the holder is left at the desired point. The stone is set free and flies into the field making a whizzing noise. That scares away the birds.

Scarecrow (Adava)

A simple contraption is a look-alike of the human being which is erected in the field. The man-like figure scares the birds and they do not come near the field. Most frequently, a stick is put on the earth and a black coloured pot is put on top of it.

Tin Box

Farmers put stones in tin boxes and beat them like a drum to make noise which scares away the birds. In another variation a string is tied to a stick in a manner that a person sitting on a charpoy can beat the drum while sitting.

Shouting

The farmer simply cries out aloud to scare the birds.

Cassette

The plastic tape of an audiocassette is tied between two trees or sticks. The tape makes a whizzing noise and drives away the birds.

Plastic Flags

A line of plastic strips or sheets is tied to sticks planted like a flag in the field. The flags make a noise in the wind and scare away the birds.

Fire Crackers

Firecrackers are set off to scare the birds. This also drives away wild animals like the Nilgai. They are very popular nowadays as they create loud noise and spark, which scare the birds away for some time.

It will be seen from the above that the farmers have smoothly synthesised the traditional and modern methods. The scarecrow has been modified to use plastic flags. Cassettes are being used instead of tin boxes to make noise. Firecrackers have replaced or supplemented the catapult. The demarcating boundary between the traditional and the modern in this case is indeed difficult to make out.

Effect

- Crop yield is better.
- Polythene is non-biodegradable. If polythene bags fall to the ground and get buried they could harm the soil. The cattle sometimes eat these bags and their intestines get choked and they die a very painful death.
- Stones shot by *Gafni* could hit humans causing serious injuries.

- All members of the farmer's family help in driving away the birds. They even give up household chores to do this because it is more important. However, women bear much of this responsibility.
- Neighbouring farmers are also benefited as birds fly away.

Cost

There is little cash cost involved. Most materials are recycled waste like used tin boxes and discarded audiocassettes. The main cost involved is that of labour of the farmer and the members of his family.

ANALYSIS

The use of traditional and modern techniques is evenly balanced. But this means that the modern techniques have made much inroads.

Survey

Type of Technique	*Number of respondents using*	*Reasons*
Traditional: Beating tin boxes,*Gafni, Adava*	8	Using tin boxes is cheaper. Even children can do it as a game. By using catapult one can drive away birds, which are very far. This is not possible by other methods. Modern techniques like firecrackers are dangerous. They can hurt the user and cause fire.
Modern: Firecrackers, plastic bags and cassette ribbons	7	Once plastic bags or cassette ribbons are tied then there is no need to do it again, while tin boxes have to be beaten daily. Less labour is required. Using firecrackers is also entertaining.

SUGGESTIONS

- Some new machines should be made that produce sound and scare away the birds.
- Farmers should be made aware about the environmental consequences of using plastics.

Stone being loaded in *gafni*

Gafni in operation

A simple scare crow (adava)

Tin box drum in operation

Plastic cassette tied along with barbed wire for making noise

Plastic cassette tied on make-shift pole

Plastic flads for noise

VIII
Harvesting and Storage

14

Seed Selection

Category: Potentially Viable

The best seed is chosen from the crop for use in the next season.

When is seed selection done?

Seed is chosen when the crop is ready. Seed selection is widely undertaken for Bajra and not for Moong and Moth. It appears that the variation of seed quality is not great in these crops.

Where is it done?

It is done in the field when the crop is harvested.

How is it done?

The basic technique is to select the better grains from the crop and store it separately for use as seed. Bajra is the main crop of the area. The seed of Bajra is selected according to one of the following methods.

- The best pods (*Sittas)* are chosen while the crop is ripe and still standing in the field. These are stored and seeds extracted separately. *Sittas* that are (1) longer; (2) with more hair; and (3) with thicker grain, are chosen. The farmer roams through the ripe crop and looks for the tall plants with more grain and hair.
- The *Sittas* may be selected after the crop is harvested and placed at the harvesting ground (*Lata*) but before the grain is removed. The grain is sieved and the larger grains are chosen after the crop has been harvested.
- The farmer looks for yellow and round seed and rejects black, white and unripe seeds of Bajra. Farmers may chew some grain to ascertain taste.

- Bajra is often harvested mechanically. The Drumi has two outlets of grain. The small grains—about 10 per cent of the total—fall at the bottom through a different spout. The main grains fall from the side spout. Farmers take care to use grains from the main side outlet alone for sowing.

Farmers select the plant itself to select the seed of Til. The pods of til are brittle and grains are too small to enable selection after harvesting.

Effect

The crop in the next season would be better.

Women undertake much of the work of selecting and storing the seed. This leads to an increase in their workload.

Cost

The only cost involved is that of one's own labour.

Seed bought from market is more expensive. The cost of seed of Bajra is about Rs. 30 per kg or Rs. 300 per *bigha*. The cost of Bajra grain is Rs. 5 per kg. The additional expense is Rs. 250 per *bigha*. The total income from Bajra is around Rs. 250 to Rs. 500 per *bigha*. The additional cost of seed is nearly equal to the total income from the crop. Therefore, it is uneconomical to buy Bajra seed from the market. Farmers prefer to select the better seeds themselves.

Festival

Seed is kept in the courtyard on *Akha Teej* and worshipped and then stored again.

Purchase of Seed

The alternative to using one's own seed is to buy seeds from the market. The main difficulty is that the seed that is bought may not be reliable. The farmers use the following methods to select seed from the market.

- Seek the advice and suggestions of other farmers.
- Choose the packed seed of a good company.

Exchange of Seed

Farmers may exchange seeds with one another or with relatives living in nearby villages. The seed obtained is generally reliable since it is obtained from a known source. No cost other than travel is involved since it is an exchange.

ANALYSIS

Survey

Name of Technique	*Number of respondents using*	*Reasons*
Traditional: Seed selected	15	Crop yield is more.
Traditional: Seed not selected	4	Seed has to be chosen during the peak harvest season when the pressure of work is very high. Choosing seed is a waste of time.
Modern: Seed bought from market	4	Purchase of seed from the market of either traditional or hybrid varieties—eliminates the necessity of selecting and storing seeds.

There are contrary arguments for using one's own seeds.

For self-grown seeds

- Self-grown seed is already adapted to the local climate and is expected to grow better.
- One's own seed would require as much water as required by the present crop.

The farmer's selection of his own seed can enhance these advantages.

For new seeds

- Seeds from the market may not suit the local climate.
- They might need more or less water according to their original environment.
- New seeds increase the yield by creating a 'culture shock' for the seeds and bringing out a positive adaptive response.

These advantages can be enhanced by procuring seed from a reliable source and from an area not too dissimilar from one's own.

The central question is that of creating 'optimal' change. Too much change may create excessive shock for the seed. Too little change may create stagnation. A typical solution used by the farmers is to select one's own seed every year and exchange it with others, every three years.

Technology	*Plus Points*	*Minus Points*	*Result*
Selection of own seed	Better yield	Stagnation	Positive Result. Better yield with nominal cost.
Use of own seed without selection	Less labour cost is incurred.	Low yield.	Negative result. Yield may decline over time.
Exchange of seed	Better yield	Danger from dew	Positive Result. unreliable seed combine with selection of one's own seed Combine with own seed.
Purchase seed from the market	Better quality and high yield.	Unreliable seed and low yield.	Negative result. Increase in yield may be minimal in comparison to the risk.

The ideal combination appears to be that of exchange of seed every two or three years combined with selection of one's own seed.

SUGGESTIONS

- Developing and disseminating better methods of seed selection can promote the technology. Farmers do not know which of the traits such as taller plant, bigger size, colour, shine or shape of the grain, and size of the pod are important. A study can be undertaken to select seeds according to these different parameters and the results can be disseminated.

- Fields may be selected for seed production instead of pods. A social system may be created for enabling exchange of seeds of the best fields between different villages.

- Discussions may be held with seed scientists to determine the optimum level of climate shock that is beneficial for the crops. Farmers may be educated about choosing the area and farmer with whom seed may be exchanged.

Selecting good pods of bajra for seed before harvesting

Selecting good pods in the *lata* after harvest

Winnowing seed of garlic

Hand selecting seed of garlic

15

Harvesting

Category : Stable

Harvesting of the major crop of Bajra is done in stages. First the pods are harvested so that the grain is safe from any natural event such as untimely rains or storms. Jwar is also similarly harvested in stages. Other smaller crops like Moong and Guwar are harvested in one go.

When is it done?

Crops are harvested when ripe. *Kharif* crops are cut in the month of *Kartik* while *rabi* crops are cut in the month of *Vaishakh.*

Where is it done?

It is done in the field.

How is it done?

The pods (*Sittas*) are cut first to get grain. This is done with the help of an implement called *Datri.* It is an iron sickle with a wooden handle. Usually all family members, including women, undertake this operation together. It is important to harvest the *Sittas* as soon as possible to prevent any loss. The farmer cuts the *Sittas* and puts them into the cloth hung on his back. When it gets filled the *Sittas* are transferred to a sack. This process is called *Ghutani.* Mostly women do this perhaps because it is considered to be 'light' work.

The stalks standing in the field are called *Doko.* These are later cut with a bigger *Datri.* The cut stalks are called *Kadab.* These are used as fodder for livestock.

Laah or *Adi Badi* (please see *Nidaan* for description) is also often used in harvesting.

Some crops like Guwar are cut in one go, i.e., *Sittas* and *Kadab* are not cut separately.

Stubs left in the field after harvesting of Kadab are used for grazing.

While cutting the crop the workers sing songs. One such song is given below.

Ram Banayo Kam, Balak Sivare Kam Bhaiyo

Budda Sivare Ram, Dev Bado Bhagwan Bhaiyo

(Lord Ram has made work; young ones do their work,

Old ones remember Lord Ram; The Lord is great.)

Effect

In the meanwhile if it rains between the first and second harvesting the *Kadab* becomes black and is less liked by cattle.

Festival

Ghutani of Bajra is not done in *Shraadh*. It is started in *Navratri*.

ANALYSIS

It was found that *all* the farmers were using the two-step technique of harvesting—first undertaking Ghurane and then harvesting the *Kadab*.

There is no visible alternative to the traditional mode of harvesting.

SUGGESTIONS

- The design and materials used in the *Datri* may possibly be improved to make it more efficient.

Bajra being harvested

Bajra being harvested

Stalks of bajra standing in field. Pods have been harvested

Pods of bajra stored.
Waiting for mechanical
thresher

Chillies being harvested

Bajra stalks *(dhoka)* are cut in the second round of harvesting

Bajra stalks *(dhoka)* cut and bundled for transport

16

Threshing (Lata)

Category: Declining

The separation of grain from the harvested plant is called *Lata*.

When is it done?

It is done when the crop has been harvested and brought to the threshing ground. It is done in the month of *Kartik* or October-November for *kharif* crops. It is done in the month of *Vaishakh* or April for *rabi* crops.

Where is it done?

It is done near or within the homestead at a level place in the open. A clean place—about 25 to 40 feet on each side or of that diameter—is smeared with dung. This place is also called *Lata*. The place must be in the sunlight so that the crop dries up. It must also be safe from animals. A fence of thorny plants is put around the land used for *Lata* so that cattle or animals do not harm the crop lying there.

How is it done?

Bajra

The pods of Bajra are spread on the *Lata* and camel or bullocks or donkeys are made to walk over it again and again in a circular motion. The hooves of the animals crush the *Sittas* or the crop into very small pieces and the grain is separated. One person keeps turning the crop over with a rake so that uncrushed *Sittas* come to the top.

The crushed *Sittas* are winnowed. Three people are required for this operation. One fills the crushed mixture in the flat basket. The second person slowly drops the mixture in a small stream on the ground when the wind is blowing. The grain falls at his feet while the chaff flies away to a distance. The third person uses a broom to sweep away the chaff that might settle near the mound of the grain.

Last, the grain is sieved to remove the remaining chaff. Then the clean grain is packed into sacks.

Guwar, Moong, Moth and Til

These crops are harvested at one go hence there is no prior separation of the pods. The crop is put on the *Lata*. Then it is beaten gently with a stick (*Lathi*). The pods get separated from the chaff and also get opened. The mixture of grain and chaff is winnowed to separate the grain.

Matira

Watermelons are harvested and stocked at the *Lata* before being loaded on transport for sale.

Chilies

Fresh harvested chilies are spread on the *Lata* to dry. They are packed in sacks when dry.

Effect

The *Lata* ensures that sand or gravel are not mixed with the grain. *Lata* is not made near a source of water such as a tube well or pond because chaff would fly and make the water dirty.

The minute pieces of chaff form a layer around the *Lata*. Cattle lick this layer. This tends to increase their milk output.

Cost

The cost of making a *Lata* has been estimated as follows :

Cleaning the area - 3 persons for labour @ Rs. 70 per person	Rs. 210
Two trolleys of *Murad* (hard sand) @ Rs. 300 per trolley	Rs. 600
Smearing with dung - 6 persons for labour @ Rs. 70 per person	Rs. 420
Two tanks of water @ Rs. 60 per person	Rs. 120
Total	**Rs. 1350**

Festival

The crop is worshipped on the day of Diwali. If *Lata* is done before Diwali and grain brought home it is believed that the goddess of wealth—Lakshmi—has entered the home.

On the day of Diwali, Goddess Laxmi is worshipped on the *Lata*.

On the day of *Akha Teej* the farmer prepares a small *Lata* at home. In it he places some grain and worships it. This is considered a good omen.

DRUMI

In the present times the farmer uses machines like the thresher to extract the grain from the crop instead of using the traditional *Lata* technique.

How is it done?

The *Sittas* are collected and heaped in a mound at the *Lata*. A thresher—called *Drumi*—is rented on a per hour basis. The Sittas are fed into the Drumi, which separates the grain from the chaff. Grain is packed into sacks. A thresher can separate grain from crop of Bajra obtained from 1 *bigha* of land in just ten minutes. A group of eight to ten persons are required to feed the *Drumi* and remove the grain.

Drumi is also used for Moong and other small crops. But in that case the entire plant is fed to the machine. The pods are not separated before threshing.

Effect

Grain is extracted very quickly and there is no danger of its getting wet by rain. Less vigil has to be kept over the crop to protect it from cattle/animals.

Less labour is required to be put in by family members. This gives special relief to women who do most of the threshing operations.

The use of *Drumi* deprives labour of valuable work. Agricultural labour obtains high incomes during harvesting and threshing operations. This income of theirs is reduced. But it may be compensated if there is a parallel increase in the cropped area and hence in the demand for labour in harvesting.

Some grain is lost with chaff in the Drumi. In traditional method all grain is recovered.

ANALYSIS

The cost of threshing by Drumi is much less than that by the traditional technology. The cost using the two methods for 12 bags of Bajra *Sitta* is given below.

Lata

Operation	*Calculation*	*Amount*
Threshing	2 pairs of bullocks @ Rs. 100 2 persons for labour @ Rs. 60 each	320
Winnowing	2 persons for labour @ Rs. 60 each	120
Lata	25% of total cost (Rs. 1350) for one year sufficient for 100 bags	40
Total		480

Drumi

Operation	*Calculation*	*Amount*
Threshing and Winnowing	*Drumi* rented for ½ hour @ Rs. 300 per hour 8 persons for labour @ Rs. 60 per hour	150 + 60
Loss of grain along with chaff	2% of 12 bags @ 50 kg per bag @ Rs. 5 per kg	60
Total		270

The cost of threshing by *Drumi* is about one-half of the cost incurred by *Lata*. There is additional benefit of quick threshing and the consequent increase in food security. The manual work to be undertaken by family members is also less. Therefore, *Lata* is much threatened by the *Drumi*. In fact a survey of eight farmers found that six were using *Lata* for small grains like Moong, Moth and Guwar. Not one farmer was threshing Bajra by the traditional method.

The use of *Lata* for small grains is explained by the small quantity in which they are grown. It is expensive to hire the Drumi machine for small quantities hence farmers prefer to use *Lata* for such crops.

SUGGESTIONS

The decline of *Lata* appears to be a *fiat accompli*. During field visits not one farmer was seen undertaking threshing of Bajra by *Lata*.

The use of *Lata* for small grains is, however, likely to continue till mixed cropping is practiced in dry land agriculture. Small quantities of different crops mature at different time and it is uneconomical to hire a *Drumi* machine for small quantities.

Empty lata

Bajra *sitta* (pods) in *lata* in the field

Bajra *sitta* in *lata*

Bajra *dhoka* lying in *lata* after grains have been removed

Bajra being removed from chaff by winnowing

The drumi

The drumi in operation

Whole plants of til in *lata.* Seed will be removed after they dry

Til being separated from whole plant manually. Seed falls to the ground.

Moong stacked in *lata*

Guar stacked in *lata*

Guar stacked in *lata*

Chillies in *lata*

Water melons in makeshift lata

17

Food Grain Storage (Kinhara)

Category: Stable

The traditional method of storing food grains is to make a vertical tank-like structure of mud with a spout at the bottom to remove the grains. This is called *Kinhara.*

When is it made?

Kinhara is made after the harvest of *rabi* crop in October-November.

Where is it done?

It is made inside the house in the kitchen, in the verandah or in the courtyard of the house.

How is it done?

A slab of stone is placed on two, three or more small pieces of stone. The purpose is to prevent direct contact of the *Kinhara* with the earth so as to prevent moisture from seeping into it. A vertical cylindrical container is made of mud. A small spout is made at the bottom to take out grains as and when needed. Cow dung is applied to the inside and outside of the *Kinhara* as a disinfectant. Grain is put in the *Kinhara* and it is sealed from the top.

If Kinhara is made in the open it may be covered with polythene sheet or jute cloth for added protection from rains.

Cost

Only cost of labour is involved in making a *Kinhara.* Stone too is locally available. A labour of about three to four woman days is required to make a *Kinhara* for storage of about one quintal of grains. At a wage rate of Rs. 30 per woman day, the opportunity cost of a *Kinhara* works out to Rs. 100.

Effect

Grain is stored in close proximity. The farmers say that grains remain safe from insects—it seems largely due to the disinfectant properties of cow dung. However, if stones are not properly placed at the bottom then moisture enters the Kinhara and termites affect the grains.

ANALYSIS

Kinhara is a standard fixture in every home. The modern alternative is to store grains in steel containers. A container of one-quintal capacity would cost about Rs. 800. At the prevailing interest rate of 24 per cent, the annual cost of a steel container works out to Rs. 200. This would be about double the cost of a *Kinhara*. Furthermore, a steel container is considered to be *less* safe from insects than a *Kinhara*—it seems mainly due to the use of cow dung in the latter. Another problem in steel containers is that the grain often 'sweats' in the heat. The taste is then destroyed.

As of now the *Kinhara* is more economical. The crucial factor, which may turn the tables, would be wage rate. If the wage rate rises to say Rs. 70-80 per woman day then it may become more economical to store the grain in steel containers. It is paradoxical that better income works against traditional technologies.

SUGGESTIONS

A comparative study of damage during storage in *Kinhara* and steel containers should be undertaken. Many organizations in the country are involved in promoting the use of by- products of cow such as *gojala*, cow dung, etc. Contact can be established with such organizations and the beneficial properties of cow dung may be documented. It may be examined how these properties may be enhanced.

Food grain *kinhara*

Closeup of the mouth

18

Seed Storage

Category: Potentially Viable

EARTH POTS

The farmers of Western Rajasthan generally store their own seed for sowing in the next season. Traditionally the seed is treated and then stored in earthen pots. The main danger to seed comes from termites.

When is seed stored?

Seed is stored after the harvest around the festival of Diwali in late October or early November.

Where is it done?

The seed is stored at home.

How is it done?

There are several variations in the methods of storage.

1. 25 per cent ash and 5 per cent dried leaves of Neem are mixed with the seed. Then the mixture is put in pots and closed with a lid. It is sealed with clay or cow dung. After every two-three months the seal is broken and the seed is spread in an open courtyard and exposed to sunlight. It is again put back into the pots and sealed. Fresh ash and dried Neem leaves are also added if there is any indication of spoilage of the seed.

2. Ash is taken from the hearth during the annual ceremonies for the departed souls of the family—the month of *pitra-paksha* or *shraadh*. The ash is allowed to cool in the night and the seed is mixed with it in the morning. If the ash is mixed with the seed in the evening then

vapours are produced which could spoil the seed. The mixture is put into pots and sealed.

3. *Balui Mitti*—a special quality of sand—is mixed with the seed and it is put into pots and sealed with dung.
4. Dried Neem leaves and *Balui Mitti* are all mixed with the seed and when dry the seeds are stored in pots, which are sealed with cow dung.

Cost

There is little cash cost involved in this operation.

Effect

The farmer does not have to worry about the availability of seed the next year. The crop can be sown quickly when it rains.

Women undertake the work of storing seed. They bear the responsibility of periodically opening the pots and exposing the seed to sunlight. The main cost involved, therefore, is of women's labour.

Festival

The seed is taken out and a small mound is made in the courtyard or the meeting hall on the day of *Akshaya Triteeya*—which marks the beginning of the new crop season. Some pieces of jaggery or a tumbler filled with water is placed in the centre of the mound. A plant of green onion or of *Bajra* is placed in the pot. Then family members pick up some seed in their palm and count the number of seeds—removing two at a time. If the seeds are even in number then no seed will remain. If they are odd in number then one would be left over. An odd number—or one seed remaining—is considered to be good and even is considered to be bad for the coming crop.

It is believed that if the seed is stored in sand on the first eight days following full moon or no moon then the insects do not attack it.

SACKS

The modern alternative is to store the seed in polythene sacks. BHC powder—a disinfectant—is mixed with seed to prevent attack by termites and other insects. Yet another alternative is not to store the seed at all and simply buy them from the market in the sowing season.

Effect

Storing in sacks is less laborious and easier than in pots. The seed also does not have to be opened and exposed to sunlight every few months.

Cost

The cost involved is that of insecticide worth Rs. 13 and a sack worth Rs. 20. Total cost is about Rs. 30-40.

ANALYSIS

A survey shows that 20 out of 27 farmers store their seed in the earthen pots. Two purchased seed from the market and five stored it in sacks with chemical disinfectants. This indicates that the traditional method reins supreme.

Name of Technique used	*Number of respondents using it*	*Reasons*
Traditional: Earthen Pots	20	Seed is readily available at home and it can be sown on time. No cash expenditure is involved. It is considered to be safer from termites than sacks.
Modern: Buy seed	2	Farmer does not have to worry about taking care of stored seed.
Modern: Store in sacks	5	Sack is easier to use. Requires less labour in comparison to earthen pots.

The farmers are combining modern and traditional technologies. Mula Ram of Kinsaria said that mixing chemical insecticides with ash and storing seed has proven beneficial.

SUGGESTIONS

A comparative study of the efficacy of different traditional and modern methods should be made. No reliable information is available, for example, of the different effects of neem leaves, ash, and ash from *shraadh*, gojala and *Balui Mitti*. The results of this study should be disseminated among the farmers.

Different technologies may be suitable for different seeds. It was reported that Moth, Moong and Chola must be stored in earthen pots while Guwar, Bajra and Jowar could be stored in sacks. This has to be studied more scientifically.

The effect of the use of plastic and jute bags has also to be studied.

Earthen pot with seed. Mouth has been sealed with dung

Series of earthen pots with seed stored

Another earthen pot with seed

Modern method. Seed stored in sacks in *pucca* room

IX

Trees and Grasses

19

Silvi-Agriculture

Category: Potentially Viable

The traditional system of crop agriculture was to protect trees that may grow in the middle of the field.

When is it done?

Trees are protected during ploughing. The ploughshare steers his plough around the trees and plants.

Where is it done?

Trees are protected in fields.

How is it done?

Trees of Khejri, Bordi, Babool, Kumatiya, Kair, Ber and Rohida grow spontaneously in the field through natural spread of seed. The farmer takes care that they are not uprooted during ploughing in particular. He only takes the lopping and does not cut down the tree itself.

Effect

Positive effects

- Prevents soil from being carried away by wind.
- Leaves falling from the trees make the soil fertile.
- Prevents erosion of soil by water. Trees provide stability to the soil during heavy showers, which cause flow of water in the fields.
- Protect the crop from hot winds—*Jhala*—which can scorch the crop.
- Trees are environment-friendly. They bring more rain to the region.
- Provide fuelwood for the hearth.

- Provide branches for making *Baad* around the field.
- Obtaining food from them for the home or selling the same generates income. Sangri from Khejri tree and Ber, Kumatiya and Kair from the respective trees are notable in this respect.
- Many trees like Khejri provide fodder in the form of edible leaves.
- Livestock sit in the shade during summers. Their dung acts as manure.
- Provide protection from dust storms.

Negative effects

- Trees cut into cropland and the yield of food grains is less. However, Khejri and Bordi are crop-friendly and they do not affect the crop yields much. Other trees adversely affect crops as they draw upon the nutrients and moisture in the soil.
- Birds come to live in trees and they cause much harm to the crop.

Cost

Most trees existing in the fields have grown on their own. No cost was incurred in growing them.

Festival

Trees are worshipped on certain festivals like Gangour.

TREE PLANTATION

The farmer may plant trees in the middle of his field. Some farmers are now planting fruit trees.

When is it done?

Trees are planted in the rainy season. They are protected during ploughing.

Where is it done ?

Farmers plant trees in their fields.

How is it done?

The farmer buys saplings from the nurseries. A hole about two feet deep is dug in the field and the sapling is planted. A protective hedge of *Chhaladiyas* or sticks is put around the plants for protection—especially from animals. Plants have to be watered during summers—up to five or six times—for the first one or two years.

Cost

About 30 plants can be planted in 1 *bigha* field. The expenditure on planting one tree is as follows:

Cost of one plant	Rs. 2
Cost of making a fence of *Chhaladiyas*	Rs. 5
Cost of digging a hole	Rs. 5
Cost of organic manure and water	Rs. 3
Miscellaneous costs, e.g., insecticide for protection	Rs. 5
Watering eight times in the first two years @ Rs 2 per watering	Rs. 16
Total	Rs. 36

Effect

The effect is same as in traditional method of protecting trees. There may be greater income if fruit trees are planted.

ANALYSIS

A survey showed that most farmers continue to protect the trees that grow spontaneously. However, nearly one-fourth of the farmers expressed disinterest in protecting the trees. The reasons have been enumerated under 'Negative Effects' above. The main resistance was on account of loss of crops by use of land, the trees attracting birds and their drawing upon the earth's nutrients. No farmer had planted trees among those surveyed. However, some examples of tree plantation have been noted. But these appear to be rare and more in the nature of an exception.

The central question is to evaluate the costs and returns from the trees. Bachan Singh of Digadi made an assessment of income from various trees as follows.

The average earning per tree is about Rs. 13. It is important to note that the income above has been estimated at purchase price—what the farmer would pay in cash if he were to purchase these things. But a cash market for these produce is nonexistent. The actual value of this income should be reckoned with reference to the realizable price, which may be about one-half of the prices mentioned above. The average from a tree may, therefore, be considered in the range of Rs. 6-13.

Survey

	Number of respondents involved	*Reasons*
Traditional: Protecting trees in field	17	All positive effects enumerated above.
Modern: Uprooting or otherwise not protecting trees in the field	6	Do not have time to take care of trees. Birds eat young plants and these die. There is shortage of water and water is not available for watering the trees.
Modern: Plant Trees	0	

Item	*Arana*	*Hingota*	*Kair*	*Ber*	*Kheep*
Number of trees	**250**	**100**	**50**	**15**	**60**
Fuel wood	800	500	200		
Timber for making *Kannera* (wooden box for keeping household goods)	1000	200			
Chhajai (sticks for house construction)	200			90	500
Sticks for construction offence	100	100	100	200	
Wood for making wood charcoal		100			
Timber for house construction			100		
Fruit				No com-mercial value	
Sticks for making ropes					200
Total	**2100**	**800**	**500**	**390**	**700**
Income per plant per year	8	8	10	26	12

The crucial question is how much crop is lost by one such full-grown tree. Babu Ram of Rampura estimated that the loss from one tree from land and nutrients was about 2 kg of grains valued at Rs. 10. There is further loss form birds, shade and loss of nutrients.

In the assessment of Babu Ram of Rampura the loss was greater. He said, "One tree per *bigha* is sufficient. More than that is not required. I do not want to save more small trees because there are enough trees." In fact, he had explicitly ordered the tractor driver *not* to save the trees and to see that they are uprooted.

On the other hand Dunga Ram of Piplad had planted a large number of Amla trees in his field and was intercropping it with Bajra. He had also saved the Khejri plants that had come up spontaneously. But this should be seen as a shift from agriculture to fruit orchard because in due course no agriculture would be possible under the trees.

It appears that the farmers believe that the returns from crop agriculture are greater than from trees. Their foremost concern appears to be that of food security.

The problem does not appear to be that of protecting trees during ploughing by tractors. Tractor drivers said that they could easily save the trees if they wanted by veering around them.

It seems that the shift from draught animals to tractors has had the effect of changing the 'limiting factor' in food grain cultivation. The limiting factor with draught animals was their number. Land was plentiful. A farmer could cultivate only so much land as the number of draught animals he may have had. In that case veering around a tree and leaving some land for it to grow would *add* to the total income. The cultivated cropland would remain at the maximum level while there would be additional income from the tree.

The limiting factor with tractor cultivation has become land itself. It is possible to cultivate all the land owned by a farmer by hiring a tractor. If he now leaves a tree, it cuts into the total area cropped. Therefore, he is no longer interested in protecting the trees.

SUGGESTIONS

- Find crop-friendly species like Khejri and Bordi which can be planted. Bajra grows well under these two trees. Planting of such crop-friendly trees may be possible. Farmers are under the wrong impression that Khejri grows spontaneously through the spread of rootstock and cannot be planted. Scientists at CAZRI, however, indicated that the spontaneous growth of Khejri plants occurs not through rootstock but through seeds. The seed of Khejri can be fed to the goats and the dropping containing softened seeds can be sown in the fields. This technology has to be tested and also propagated.

- Undertake research on the optimal spacing between trees and between lines of these trees. Planting in lines can enable cultivation by tractors. Farmers have to be made aware that it would no longer be adequate to rely on the spontaneous growth of these trees because they obstruct cultivation by tractor. The need to consciously plant these trees has to be communicated.
- Find trees which give high value returns so as to make them more economical in comparison to crop cultivation.

Hingoti tree in field

Ber pala in field

Pomergranate tree in field

Farmer showing khejri tree that he has saved in the field

Farmer showing khejri tree that has grown by itself on the field border

Plant of aakh saved despite ploughing by tractor

Ber tree saved in field

Amla trees planted as orchad

20
Tree Lopping (Loonkh Katana)

Category: Stable

Loonkh cutting means lopping of branches of trees.

When is it done?

Lopping of the trees is done after the harvest of crop, i.e., after November.

Where is it done?

Lopping is done of various trees, Khejri in particular, in the field.

How is it done?

An axe is used for lopping. A person climbs a tree and seated on a strong branch he cuts other branches. The cattle may be allowed to eat the leaves there. Alternatively, the branches are collected and taken home. Another method is that after cutting the branches they are collected at one place and allowed to lie there for 15-20 days. The branches dry. The dry leaves fall down and are fed to the cattle while the branches are used for various purposes.

A smaller implement called *Datri* is used for cutting small branches for the purpose of fodder. However, 'lopping' is done only by axe when larger branches are cut. The re-growth is better when larger branches are lopped by axe.

It is considered inauspicious to cut all branches. Hence a few branches are left for the future growth of the trees. If branches are not cut the tree does not grow and spread. If lopped the tree grows better.

Effect

The leaves of the cut branches—*loonkh*—are fed to livestock. They can also be sold. The dried branches can be used as fuelwood or to build *Baad* around the field.

Lopping of branches of Khejri makes the branches and the tree grow better.

Cost

The only cash cost involved is that of an axe—about Rs. 50. Rest of the work is done by one's own labour. One man can do lopping of one Khejri tree in one day. The cost is his wage of Rs. 60.

Income is as follows:

Leaves	Rs. 30
Branches	Rs. 20
Total	**Rs. 50**

The direct income is Rs. 50 against the prevailing wage rate of Rs. 60. Therefore, lopping is done when other agricultural works have been completed and opportunity cost of labour is low.

ANALYSIS

A survey indicated that every respondent lopped his trees. Lopping is an essential operation for better re-growth of the trees. There is no alternative.

SUGGESTIONS

Undertake a detailed comparative study of lopping by *Datri* and axe. Study the regrowth of the tree. This may be done separately for Khejri and other trees like Babul. If it is found that the loss from axe is greater than the saving from labour then the farmers can be educated about this fact and encouraged to use *Datri* alone.

Lopping under operatrion

Lopped branches waiting to be transported. Leaves have been eaten by small livestock

Livestock Eating leaves from lopped branches

Leaves collected from lopped branches by allowing them to bry and fall off. here being fed to small animals

Livestock Eating leaves from lopped branches

21
Harvest and Storage of Ber Stalks (Pai)

Category: Declining

Ber is a small shrub that grows profusely in the wild as well as in the fields. Children enjoy its small red berries. The leaves—called *Pala*—are good fodder for livestock. The thorny stalk is excellent material for border fencing. The Ber stalks are stored in heaps called *Pai.*

When is storing done?

The storage in heaps is done in the month of October (*Kartik*) after the rains and after the crops have been harvested.

Where is it done?

The harvesting and storage is done in the field where there are bushes of Ber.

How is it done?

Two tools—the *Bai* and *Kulhadi* (axe)—are used together for harvesting the *pala*. The farmer presses the bush down with the help of the *Bai* and cuts it with the *Kulhadi*. Then he carries the cut *Pala* on the top of the *Bai* to a particular spot where the earth is hard and spreads the stalks side by side. The stalks get dried and the leaves fall to the ground. The stalks are then again picked up with the *Bai* and now heaped one over the other in the form of a mound. This is called the *Pai*.

Farmers often tie an additional black cloth on their waist while cutting *Pai* to protect their apparel from damage due to the friction of the *Bai*.

Effect

This operation provides fodder for livestock. This fodder is much liked by the small animals like goat. It is supposed to be very nutritious for them. The family can sell Pala and Pai and get additional income.

The process also clears the field of the bushes. These grow back again next year, as the roots are deliberately not removed.

Bushes of Ber make the soil fertile. The leaves that drop off act as manure. The roots of the bushes also trap water and help retain moisture. They also prevent erosion of soil by wind and water. Farmers indicate that Ber plants do not compete much with crops. The shallow roots of Ber are uprooted during ploughing and crops are sown. Ber plants grow again in the midst of the crop.

Women have less to labour for getting feed for cattle, as *Pala* is available.

Festival

Fence is made on festival days in homes with the *Pai.*

ANALYSIS

The technique has no alternative. The use of *tawi* reduces the yield of *Pala* in one year but it grows back again in the next. But after the *tawi* is used three times the major roots called *Goonth* are destroyed. Then regrowth does not take place. Farmers typically use the *tawi* every three years. Thus ber is destroyed after nine or ten years. The availability of *Ber* has already declined substantially. Nowadays *Baad* is being made with plants like *Aakh, Phog,* etc.

The field has to be cleared for ploughing. The stalks are excellent material for fencing. The use of barbed wire may reduce the utility of the *Pala.* But this question is really related to the wage rate. If the wage rate is high and the farmer can earn cash incomes from work as labourers then he may give up making fences of *Pala.* This appears to be a remote possibility for the present.

A survey showed that all farmers but one were harvesting the *Pala.* The lone exception did not have any Ber plants in his field.

SUGGESTIONS

Experiment of planting Ber on bunds of *Khadins* can be made. Ber plants would impart stability to the structure and also provide income. This requires a change in the mindset. Traditionally Ber grows spontaneously and farmers merely harvest the *Pala.* Now efforts have to be made to plant it on lands that are not subject to cultivation by *tawi.*

Survey

	Number of respondents involved	*Reasons*
Cutting *Pai* and *Pala*	17	*Pala* is obtained for fodder and *Pai* for making fences. There is less erosion of soil due to presence of the bushes. The soil also remains fertile.
Pala and *Pai* are not cut	1	There are no bushes in the field.

Ber *pala* being cut

Ber *pela* being carried for initial stacking. Leaves will be allowed to drop off and be collected

Stack of ber leaves being covered

Ber stacks being stacked after leaves have been collected

22
Storage of Wood (Pindara)

Category: Potentially Viable

Pindara is a shelter made to store fuelwood so that it remains safe from termites and rain.

When is it done?

It is done in summers and winters when the men and women are free from work in the fields. In the summers it is done around *Holi* festival and in winters after *Diwali*.

Where is it done?

It is done near the home so that care of the stored fuel can be taken care of from time to time.

How is it done?

The place where the *Pindara* is to be made is first cleaned and then stones are laid and ash sprinkled to protect the stored wood from termites. Then the wood to be stored is placed in a heap. The wood is laid in the same direction and as closely as possible to reduce the water that may enter the heap. The height is generally six feet. The whole structure is pyramidal in shape, being broad at the bottom and tapering at the top.

Cost

The family members themselves stack the wood and make the *Pindara*. No cash cost is incurred.

Effect

The wood is wetted less by the rain. In times of rain wood can be taken out from the *Pindara* and used. When family members are busy in the fields and have no time to go out and collect wood for fuel or prepare dung cakes, then the fuel from the *Pindara* can be used. The fuel can be used in marriages and other functions.

Women do not have to wander about in search of wood for fuel.

The land where the *Pindara* is built becomes fertile. More fuel can be stored in less space.

Wood in Open

Many farmers just lay the cut wood in heaps in the open.

Effect

The wood occupies a lot of space. It is easily attacked by termites and destroyed. It gets wet wholly during the rains.

LPG

The modern alternative is to use LPG fuel instead of fuelwood.

Cost

A capital expense of Rs. 3,000 is required as security deposit. The cost of refilling a cylinder is Rs. 300.

Effect

It is easier to cook food on stoves fueled by LPG or kerosene. Little smoke is produced. It is more comfortable for the women to use.

ANALYSIS

All farmers using fuelwood are making *Pindara* for storage. Only those who have very little wood do not make *Pindara*. Therefore, the technology of *Pindara* is well entrenched *if* fuelwood continues to be used. The main 'threat' comes from modern fuels like LPG and kerosene. Increasing use of these fuels will overtake fuel will cease wood and, therefore, the making of *Pindara*

SUGGESTIONS

The problem has to be looked at from a larger ecological perspective. Use of wood as fuel has severe ecological implications in view of declining availability of biomass and forests. On the other hand, blind reliance on fossil fuels like kerosene and LPG is also troublesome because of the dependence it builds on imports. The correct direction would be to move towards a sustainable fuel policy based on renewable energy. This would imply cultivation of biomass farms. The technology of Pindara has to be dovetailed into that larger strategy. Use of dung as fuel should also be assessed in this backdrop. That would impinge upon technologies such as Thepadi storage.

Survey

Technique used	*Number of respondents using it*	*Reasons*
Traditional: Pindara	19	Fuelwood does not get spoilt. Use of kerosene and LPG is costly and the same are not available on time. *Chapattis* of Bajra are prepared better on fire made from dung cakes than on gas stoves. The ash left is used as insecticide in fields.
Traditional : Wood in open	3	There is not enough wood to make a *Pindara*.
Modern: LPG	3	The gas stoves do not produce smoke. They are easier to use. Making dung cakes and *Pindara* is difficult and involves a lot of effort.

SUGGESTIONS

Develop a rural fuel policy with stress on renewable sources of energy. Develop fuel farms within this strategy—select species suitable for Western Rajasthan. Tree species like *Acacia Tortilis* or the like may provide a solution. Dovetail storage technology in tune with this policy.

Pindara may be made of stalks of wild plants like *Aakh.* These plants grow profusely on common grazing lands. They are presently not used much for fuel. Similarly the stalks of castor that is planted to provide protection to chilies from wind can be stored in a *Pindara.*

Pindara of fuelwood

Pindara smeared with dung

23

Tree Produce (Ber, Kair and Kumatiya)

Category: Potentially Viable

The fruit of Ber is sweet and is relished by the people. Kumatiya fruit is used as a vegetable and it is very nutritious. Kair fruit is used as a vegetable and for making pickles.

When are the fruits picked ?

Ber and Kumatiya fruits are picked after the festival of Diwali, i.e., in October or November.

Kair fruits are picked in April. These are picked unripe and pickled.

Where is it collected from?

Fruit is collected in *Oran, Gochar*, roadside or one's own field—wherever the trees are growing.

How is it done?

Ber and Kair bushes have thorns, which can hurt. A long bamboo stick is used to shake the bush. The ripe fruit fall to the ground and are collected. Before doing this the earth under the tree or bush is cleaned to remove thorns that might prick when the fallen fruit is picked.

Ber grows in two forms—shrub and tree. The shrub grows widely in the field. As a practice these shrubs are cut after the crop is harvested. The sticks are used for fence and the leaves for fodder. This cutting is not destructive as the shrubs reappear again in the next season. Children pick fruit from the shrubs as soon as they are ripe.

The Kumatiya trees are similarly beaten with a long bamboo and the ripe fruit fall to ground from where they are picked.

Kair fruit is picked one by one. The bushes are not harmed.

Effect

Children pick Ber fruit for eating. Ber fruit is sweet and is liked by most people. It is dipped in water in summers and then eaten. This prevents the harmful effect of *loo*—the hot winds—that can make one sick.

Leaves of Ber and Kumatiya are nutritious for the goats and sheep and are relished by these animals.

Kumatiya is used as a vegetable. There are hardly any green vegetables in villages and Kumatiya address that problem.

Mostly women do the work of picking the fruit.

Cost

Ber

Cost	Two persons for labour (children or women) @ Rs. 35 per head	Rs. 70
Income	10 kg Ber @ Rs. 10 per kg	Rs. 100
Net Income		Rs. 30
Net Income per manday		Rs. 15

Kumatiya

Cost	Three persons for labour (children or women) @ Rs 35 per head	Rs. 105
Income	5 kg Kumatiya @Rs 80 per kg	Rs. 400
Net Income		Rs. 295
Net Income per manday		Rs 100

Kair

Cost	One person for labour (child or women) @ Rs. 35 per head	Rs. 35
Income	5 kg Kair @ Rs. 15 per kg	Rs. 75
Net Income		Rs. 40
Net Income per manday		Rs. 20

The net income from Ber and Kair is about Rs. 20 per man-day. Both these plants are found aplenty. There is little 'rent' involved. Kumatiya, on the other hand, is a scarce tree. The high income is due to the high price or scarcity of the tree.

Destructive Picking

Nowadays branches laden with fruit are cut with a tool called *Tadi.* They are brought home and the fruits are picked at leisure. This results in destruction of Kair bushes.

People do not have to stand in the sun and pick the fruit. They can do so comfortably at home.

ANALYSIS

Survey

	Number of respondents involved	*Reasons*
Traditional: Picking Kair fruit at the site with hands	4	Protection of these bushes means more bushes and a pure environment.
Modern: Cutting branches of Kair	4	The branches can be taken to a comfort able place and the fruit picked from them at leisure. The hands of the picker are more protected from thorns. Even children can contribute to picking Kair at home. The fruit is picked quicker.

The shift from constructive to destructive harvesting of Kair is indicative of the higher level of wages. The returns from picking are small in relation to the time involved.

Trees of Ber and Kumatiya are not generally harmed. Kumatiya is a scarce commodity hence its use is likely to continue. But farmers may adopt destructive harvesting of Ber.

The farmers cannot be asked to continue with hand picking which involved a low income. The task is to increase the yield of picking without harming the tree.

SUGGESTIONS

Make social rules for harvesting of these forest products. The use of community resources suffers from the problem of grazing dilemma. Every person would like to himself violate the social norm while expecting others to adhere to it. This problem can only be tackled by enforcing social rules. In some villages the people get together and fix a particular day on which everyone in the village would harvest these fruit products. There is less chance of destructive harvesting when the work is done under communal gaze.

Ber tree with fruit

Children have collected ber fruit from tree

Kair tree

Closeup of kair fruit in the tree

Woman harvesting kumatia

Kumatia pods

24

Protecting Small Trees (Jhopa)

Category : Stable

A shelter called *Jhopa* is made over small plants in order to protect them from the strong sun in summers, the cold in winters, the winds and stray or domesticated animals. The shelter also helps retain moisture in summers.

When is it built?

The *Jhopa* is built after the plants have been planted.

Where is it done?

It is built over plants are planted in fields, near home and near Tanka.

How is it done?

Earth is dug at a radius of about two feet around the plant and sticks are planted. The sticks are about two times taller than the plants. The sticks are bunched together at the top and tied. Over this skeleton grasses like Sinia or Kheep are tied. When the shelters are no longer needed and the plants have grown into trees the bushes used to make the *Jhopa* are thrown into the field and they become manure.

Cost

Making of a *Jhopa* involves only locally available material and labour. Farmers estimate the cost of one *Jhopa* to be Rs. 5.

Effect

Moisture is retained in the soil due to the shade of the shelter and there is less transpiration hence the plant has a greater chance of survival. It has to be watered less frequently.

In winters the *Jhopa* protect the plants from extreme cold.

For the soil too the technique is beneficial. The leaves fall to the earth and are not carried away by wind. This makes the soil fertile. There is less soil erosion.

Trees are also protected from animals that could eat and destroy them. The shelters are covered with thorny bushes for this purpose.

Once shelters have been built over the plants the family members do not have to worry about their safety or keep constant vigil. Generally women take care of the plants. Then they become free to do other chores.

Festival

On the day of *Teej* and other festivals trees like Peepal and Khejri are worshipped. Women and children enjoy swings hanging from trees on the day of *Teej*.

WIRE NETTING

Some farmers make fences of wire, barbed wire or wire netting around the plants for protection.

When is it done?

Fence is put around the plants when they are planted.

Where is it done?

Fence is put where the plants are planted.

How is it done?

The iron wire or wire netting is bought from the market. These are available in rolls. The right size is cut and tying the wire in a circular shape forms a cylindrical fence. The earth around the plant is dug and the fence is put. Such a fence can be up to five feet high.

Cost

The cost of iron fence around each plant is estimated to be Rs. 60.

Effect

The plants are properly protected from animals. A fence is stronger than a *Jhopa* and the animals cannot reach the plants inside. But there are disadvantages of the technique too. Firstly the iron fence gets rusted. The plants are not covered or protected by shade therefore they need more water. They are not protected from the summer heat and winter cold.

However, the fence is beneficial for the family members and they do not have to keep vigil over the plants to protect them from animals.

ANALYSIS

There is virtually no competition to *Jhopa* from fence technology. Of thirty persons surveyed only two were using fence for protection of the plants. *Jhopa* provides all round protection from wind, heat, cold and animals. Its only disadvantage is that it needs to be repaired frequently and is not very durable. Three farmers preferred not to provide any protection. Perhaps they were growing a few trees in a protected situation within the homestead. But the *Jhopa* is a entrenched well technology.

Survey

Technique Used	*Number of respondents using it*	*Reasons*
Traditional: *Jhopa*	25	There is less need to water the plants. The plants are protected from strong sunlight in summers and extreme cold in winters. More moisture is retained in the soil around and in the plant. Plants are protected from strong winds. The cash cost is less.
Modern: Iron fence	2	It is long lasting and need not be repaired frequently. It protects the plant better from animals.
Modern : No protection	3	Plants can grow without protection hence there is no need to make a *Jhopa*. Do not have time to make the *Jhopa* over the plants. *Jhopa* requires a lot of labour.

Grass jhopa

Grass jhopa

25

Wild Grasses (Kheep and Sinia)

Category : Declining

Kheep and Sinia are the two more widely found wild grasses in Western Rajasthan. These are used to make huts, ropes, cots, shelters for cattle and fences around homes and fields. These are also used for *Kanabandi* and *Jhoor* around the fields. Ropes made from these are used to make *Jhopa* and for tying *Kalar.* Ropes made from Sinia are used to make cots.

When are they cut?

The grasses are cut in the month of November while they are not fully matured and have not become hard. Once ripe, they become useless.

Where is it done?

They are cut around the village and in fields where Sinia, Kheep and other grasses grow.

How is it done?

The farmer uses an iron axe to do the harvesting. Grasses are cut and placed flat or one upon the other. Some weight is put on top to press and straighten them. Then they are used when needed.

The roots of the plants are not destroyed so the plants can grow again. The roots prevent erosion of soil.

Kheep is soaked in water and then buried in the earth. It becomes soft. Next day it is taken out and rope is made out of it. Ropes made from Sinia are used to make cots. People use earthen pots such as *Kalsi* in the area. A round ring of glasses called *Arai* is put on the floor to prevent the pots from tilting. This *Arai* is made from these grasses. The grasses are used to make shelters for cattle. These shelters last for up to three years. Farmers make *Jhopa* (for protecting small tree plants) or huts with these grasses. The grasses are tied to a structure of sticks with ropes to make a roof.

Effect

The root of grasses trap water and keep the soil moist.

Many grasses such as Kheep and Sinia serve as fodder for sheep and goats.

Festival

Roof of huts are built on festival days.

ANALYSIS

Houses made of cement and mortar are becoming popular hence the use of grasses has decreased. Huts made of grasses have to be repaired each year hence they are not in fashion any longer.

Plastic ropes are easily available in the market. For cots too the people prefer plastic ropes that are more durable.

The use of grasses is labour intensive. The central question appears to be that of availability of employment and the wage rate. If employment is not available then farmers would use their time harvesting the grasses and making huts from the same as the opportunity cost of labour is low. If the wage rate is high then they would abandon the use of these grasses. The use of tractor makes it possible to cultivate the land and simply remove the sticks or burn the grasses which are uprooted. The continued use of grasses depends upon the wage rate. This is the main threat to this technology.

	Number of respondents	*Reasons*
Making of grasses	14	They are cheaper. The huts made from these grasses remain warm in winter and cool in summers.
Not using grasses	Nil	

It will be seen that all farmers are using grasses at present. This indicates that at present wage rates are low and there is no immediate threat to these technologies.

SUGGESTIONS

Scientific study of these plants may be made to see if some chemicals can be extracted from them.

Sinia

Kheep

Masur

Bui

Bekaria

Jinjani

Grass uprooted during ploughing being collected

Grass being cut

Grass being stacked

Grass being stacked

Sinia grass being wetted and buried in sand for softening before making rope

Making *arai* (pot holder) from Sinia grass

Arai made from sinia

Rope being made from sinia

X

Field Management

26

Fencing (Baad)

Category: Potentially Viable

One difficulty in farming in Western Rajasthan is to protect the crop from wild animals—the Nilgai, in particular, human beings and insects. The farmers make an obstruction on the border figures called a *baad* for the purpose. Immense variation is found in the types of such *Baad*. The accompanying for the purpose give some idea of the same.

- **No baad**. Farmers may leave the field unprotected where other cultivated fields lie adjacent to their own.
- **Dry bushes**. Dry bushes and sticks may be inserted directly into the earth or on a mud wall.
- **Baad with ditch.** A ditch may be dug parallel to the *Baad*. This is especially useful in preventing the attack of Katara insect on bajra crop. This insect crawls from one field to another. It falls into the ditch and is unable to get out. When it is trapped the farmer kills it by burying it in the earth. The farmer digs holes in the field to trap the insects and kills them by burying them if Katara already infests the field.
- **Baad with live bushes**. I ive bushes may be planted on the mud wall. A variety of plants are used for the purpose. More popular plants are Thor, Aakh and Babul. The plants provide stability and require less upkeep but they also occupy more land.
- **Only stone**. Stone slabs or bricks may be laid to make a wall.
- **Stone with live bushes.** Cactus or other plants may be grown parallel to the stone wall for added protection.

- **Stone with barbed wire or dry bushes.** Stone wall may be reinforced with a barbed wire or dry thorny bushes or both.
- **Trees along with dry bushes.**
- **Trees along with mud wall and dry bushes.**
- **Khejri tree growing spontaneously is protected to double up as Baad.**
- **Trees planted with barbed wire**. Trees are used instead of stone pillars to hold the barbed wire in position.
- **Barbed wire**. Only barbed wire is laid on stone pillars.
- **Dry bushes with barbed wire**. Barbed wire is held in position by stone pillars and reinforced with dry bushes.
- **Moonja with mud wall.** Moonja grass has not been used traditionally for *baad* but is now being introduced. The grass provides material for thatched roofing.

Indeed, it is difficult to determine which of the above techniques are 'traditional' and which are 'modern'. Stones, barbed wire and trees may generally be considered to be 'modern' because they have been recently introduced and are less pervasive. We give below detailed analysis of some of these alternative *Baad* below.

BAAD WITH BUSHES

In this technique bushes and thorny branches are used to form *Baad* around the border of the field. This is done mainly to protect the crop from stray animals and winds, to prevent soil erosion and to make boundaries of fields permanent. A high *Baad* protects the crop from *Jholaa*—a hot wind— which dries the soft grain and stunts its growth. Often a ditch is made parallel to the *Baad*. The insect Katara falls into the ditch and is not able to climb up.

When is it made?

Baad is made when the seed has germinated and new saplings have appeared in the field typically in the month of *Ashad,* i.e., June-July. The *Baad* made from dry bushes or mud is prone to decay. Hence the farmers make it as late as possible but before the crops become big enough to attract wild animals.

Where is it done?

The *Baad* is made all round the field on its boundary.

How is it done?

The earth is dug and dry sticks are planted. Mostly thorny sticks of Ber and Babool are used.

Effect

The *Baad* is beneficial in many ways.

- The soil is not blown away by wind or carried away by water.
- Rainwater is better absorbed by the soil as it is trapped in the field and does not flow away.
- Stray livestock and wild animals cannot enter the field and harm the crop by eating or trampling it.
- Erection of a *Baad* does away with the necessity of the family members staying awake through the night to keep vigil against animals—the Nilgai in particular.
- The neighbour is saved the trouble of making a *Baad* on his side of the field. More often a heavy *Baad* is built along the sides of the roads, grazing lands, Oran or other areas where stray animals may come freely.
- It acts as a marker for delimitation of the fields.

Cost

The cost of making *Baad* on one *bigha* land is estimated as follows:

10 persons for labour for digging @ Rs. 100 per person	Rs. 1000
Cost of dry sticks	Rs. 200
2 persons for labour for making the fence @ Rs. 80 per person	Rs. 160
Total	**Rs. 1360**

BAAD WITH TREES

Trees and bushes can also be planted on the borders of fields for protection. Trees help reduce soil erosion by slowing the velocity of wind. In summers cattle get shade from the trees and their droppings make the soil fertile. Trees are lopped for cattle feed and also provide fuelwood for the hearth.

Stray examples of trees planted along the *Baad* are found in the area. It is, however, not the prevailing practice. The main reason for not using this technique is that trees encroach upon valuable cultivable land. This same problem is encountered in respect to ploughing. Farmers will deliberately uproot small trees because they do not want to lose crop yield. The farmers find the returns from crop cultivation to be greater than those from tree plantation. Hence few orchards are seen in the area.

Trees, however, may be planted where a road, grazing land or other common land lies on the other side of the road. The land lost on the other side of the *Baad* would not be that of the farmer.

The major factor is that of labour. *Baad* making is a labour-intensive operation. It requires remaking every year. Farmers who are aged or who have alternate source of employment in the urban areas want to reduce the investment of their time in agricultural operation. Such farmers are more inclined toward tree plantation.

One modern variation of this technique is to plant trees of Kheri on the *Baad*. This tree requires little water to survive. Seeds can be directly sown on the *Baad*. It does not require growing saplings and transplantation. Gum can be extracted from Kheri trees. It is sold @ Rs. 150 per kg and fetches extra income for the family. One Kheri tree produces 250g of gum.

Yet another variation is to plant stubs of Moonja grass instead of trees. This grass does not require any watering to grow. The grass is used to make thatched roofs.

When is it done ?

Planting of trees is done in the rainy season.

Where is it done?

It is done around the field.

How is it done?

Different trees like Aranna, Hingoti, Khipraa, Kair, Kankeda and Ber are planted along the boundary of the field with a gap of five feet. Plants are brought from nurseries. First the earth is dug and then the young saplings are planted. They are watered once when planting is done. And then they have to be watered occasionally till the roots go deep enough to sustain the tree. The main difficulty is of sustaining the trees in the summer of the first year when they have to be given water about three times.

Cost

The cost of *Baad* with tree plantation on one *bigha* is estimated as follows:

10 persons for labour for digging @ Rs. 100 per person	Rs. 1000
Plants - 100 nos. @ Rs. 2 each	Rs. 200
5 persons for labour for planting @ Rs. 80 per person	Rs. 400
Watering five times @ Rs. 400 per watering	Rs. 2000
Total	**Rs. 3600**

Effect

Trees provide a permanent protection against stray and wild animals.

Farmers recognize that tree plantation is environment-friendly and brings more rain.

Lopping of trees provides fodder for the livestock and fuelwood for the hearth.

The trees have to be watered and protected from animals. This work is undertaken by women leading to an increase of their workload.

BARBED WIRE FENCE

The modern technique being increasingly adopted is that of putting a boundary of barbed wire around the field. It requires little annual maintenance.

This technique is very frequently used in irrigated fields. Irrigated agriculture is undertaken on two crops—*kharif* and *rabi*. Dry land cultivation is undertaken only in *kharif*. In the *kharif* season almost all cultivable land is under cultivation. The herdsmen are expected to be present with their animals and see that they do not graze a standing crop. This practice comes to an end with the harvesting of the *kharif* crop—mainly Bajra—in October. Thereafter even domestic livestock are left free to graze the stubs of bajra remaining in the field. The unherded animals are ever trying to enter the lush green fields, which have been irrigated. These fields, therefore, stand like an oasis in the desert and have to withstand much attack from animals. It is here that barbed wire fences are frequently used.

When is it used?

Barbed wire is put around the field before sowing of seed and generally after rains.

Where is it done?

It is done all round the field.

How is it done?

All round the field holes are dug at a distance of 20 feet. Stand posts of stone about seven feet long are put upright. Then four lines of barbed wire are tied to make a fence. Dry bushes are tied to these barbed wires to prevent small animals like sheep from entering. In one variation trees are planted at a distance of approximately ten feet and their trunk is used as stand post instead of erection of a stone pillar.

Effect

Cattle and unwanted animals are kept out very effectively by barbed wires. However, erosion of soil by wind and loss of water is not prevented by the barbed wire fence. On the other hand, this *Baad* requires very little land.

Cost

The cost of putting barbed wire in one *bigha* field is estimated as follows:

10 persons for labour for digging @ Rs. 100 per person	Rs. 1000
20 stone posts of 7 feet each @ Rs. 4 per feet	Rs. 1680
1 quintal barbed @ Rs. 17 per kg	Rs. 1700
10 persons for labour for making fence @ Rs. 100 per person	Rs. 1000
Total	**Rs. 5380**

ANALYSIS

The farmers have become fully conversant with the modern techniques such as barbed wire fencing and planting of Kheri trees. They have made a large number of variations in the process. They have frequently merged elements of traditional technology with modern technology such as by tying dry thorny sticks between barbed wires or using trees to hold the barbed wire instead of establishing stone pillars. A survey found that a large number of farmers were still making *Baad* of dry bushes with- or without the trees. The change from traditional trees to Kheri and Moonja is slowly taking place. Only one farmer had made a *Baad* of barbed wire. This is along expected lines as barbed wire *Baad* fence is more frequently made by those few who own tubewells and engage in irrigated agriculture.

The above survey indicates that the traditional technology of bush-with-tree is holding its ground. The factors that influence the selection of technology and, at present, are arrayed in favour of bush-with-tree are as follows:

Technique used	*Number of respondents using it*	*Reasons*
Traditional: Dry Bushes	17	Protects the crop from Katara. It is cheaper to use them. Members of family can do it. There is no need of engaging outside labour. Bushes are available in one's own field. Cash expenditure is not involved. Otherwise soil and organic matter flow away with rainwater or wind. Crop is protected from hot winds.
Traditional: Trees on *Baad*	12	Planting trees prevents soil erosion. Leaves falling to the ground and provide organic manure. Fodder and fuelwood are obtained. Timber is obtained for house construction.
Modern: Kheri tree	3	Gum obtained from the trees can be sold. The tree needs less water to grow.
Modern: Moonja grass	4	Moonja can be used to make huts, brooms and rope. It can be a source of income.
Modern: Barbed wire fence	1	It is durable. There is no need to put up *Baad* of bushes again and again. It provides more secure protection. Livestock cannot enter easily.

- Location of the field. If field is adjacent to another cultivated field then a weaker *Baad* is sufficient since the neighbour would be protecting his field from animals on his side. Farmers having fields on the sides of major highways make stone *Baad*—to keep away both human beings and animals.

- Availability of grazing lands. Farmers need dry bushes for the *Baad*. It is often not possible to find adequate quantity of them from their own fields—availability of bushes from own fields has been declining due to cultivation by tractors. If the village has a large common grazing land then farmers have a greater inclination towards traditional method.

- Opportunity cost of labour. Farmers who have access to city-based wage employment are less inclined to spend time in construction of traditional *Baad*. Those who are old and unable to make *Baad* every year are more inclined to plant trees.

- Availability of city-based cash income. Those who have income from relatives residing in the city are the ones who more frequently have tubewells. They also have the money to put up *Baad* of barbed wire.

SUGGESTIONS

The traditional *Baad* is holding its position. However, it will come under increasing pressure as a larger number of rural people move to cities or engage in non-agricultural vocations. The availability of wage employment throughout the year has meant that the previously 'lean' period of summer, during which the opportunity cost of labour is small, will no longer provide time for undertaking these activities. A pro-active strategy, therefore, would be to reduce the labour involved and increase the returns. Our suggestions are as follows:

- Study techniques of establishment of Khejri tree on the *Baad*. It is clear that trees provide an ideal solution for *Baad*, especially in combination with barbed wire but they occupy considerable land area: about 10 to 15 feet land along the *Baad* is rendered unfit for cultivation. The main obstruction in the promotion of tree-based baad is this loss of cultivable land. The well-known Khejri tree—which does not eat into cultivable land— is an ideal solution. However Khejri is not successfully raised in the nursery and planted. It propagates through roots and new plants grow spontaneously. This growth has lately been hit by the use of tractors and deep ploughing, which cuts the roots that may be in the process of growing into trees. At the farm of Dungar Ram at Piplad it was found that small Khejri plants had emerged near the *Baad* itself. This experiment needs to be studied in greater detail. Might it be possible to train the Khejri roots to emerge near the *Baad*, rather than in the middle of the field?

- Study suitability of other species of trees and shrubs for planting on the *Baad*. Other tree species have to be studied for their suitability for silvi-agricultural cultivation. Trees like Kheri and shrubs like Moonja have little synergy with agriculture. Ratanjyoti—a shrub that grows in Southern Rajasthan—is suitable for planting on *Baad* and also provides oilseeds, which fetch good income. Some varieties of agave cactus are known to provide raw material for the production of alcohol and other chemicals. Thor—the cactus that is used in Western Rajasthan—does not provide such commercially valuable products. Experiments can be undertaken with other species of trees and shrubs that simultaneously provide protection and commercially valuable products.

No *baad.*
Adjoining area is also cultivated

Dry bushes as *baad*

Mud wall with dry bushes

Mud wall with dry bushes

Mud wall with dry bushes in making

Mud wall, dry bushes and ditch

Mud wall
with bush of
vilayati babul

Mud wall
with aak
plants

Mud wall
with thor

Baad with stone and live bushes (cactus)

Stone *baad*

Stone with
dry bushes

Stone with
barbed wire
and dry
dry bushes

Stone pillars
with barbed
wire

Thees with dry bushes

Trees with dry bushes

Mud *baad* with kheri seed sown on ridge

Khejri tree has grown spontaneously. It is being protected to grow and double up as *baad*

Trees with barbed wire

Baad with barbed wire

Mud wall with moonja reinforcement

27

Fallow Lands (Adav)

Category: Declining

The farmer leaves a part of his field or land uncultivated for one or more crop seasons to provide fodder for livestock and to enable the fertility of the soil to regenerate. This is called *Adav*.

When is it done?

Land is left fallow during the rains.

Where is it done?

The fields are left fallow. Often the fallow is rotated between different fields. The weakest field is left as *Adav*.

How is it done?

The farmer divides his field into two parts. The more fertile part is used for crop cultivation. The less fertile land is left as *Adav* for the cattle to graze. A fence of bushes is put around the *Adav* to mark its boundary and to prevent soil erosion.

If a farmer does not want to grow any crop at all in the season, he may leave the whole field as *Adav*.

Usually *Adav* is left for three years so that the soil regenerates substantially.

Cattle are tied on one side and grass is allowed to grow in the rest of the area for cutting at the end of the rains. This is done more frequently when the farmer has less land for leaving as *Adav*.

Effect

Adav has the following effects:

- *Adav* provides land for the livestock to graze upon—specially during the first four to six weeks of the rains when fields are cultivated and common grazing lands are already denuded from overgrazing and summer heat.

- *Adav* improves the soil fertility and gives a better crop in the next season. The soil gets rest. Incidentally, the use of *tawi* has the same effect. The topsoil is given rest by overturning.
- Droppings and dung of livestock during grazing make the soil fertile.
- *Adav* decreases erosion of soil by wind and water. The grass cover prevents loss of topsoil.
- Livestock do not enter neighbour's fields and graze in one's own land. Good relations are maintained with neighbours.
- The workload of women is reduced since generally they have the responsibility of taking the cattle for grazing.
- Women get dung from *Adav* for making *thepadi* for fuel.
- A particular grass—Bhroot—grows in the *Adav.* This grass is crushed and *chapatti* made out of it. This practice is receding now a days.

The negative effect of *Adav* is that the farmer has to forego crop cultivation and, therefore, the food security of the family is impaired.

Cost

The cost of *Adav* is mainly in the opportunity cost of loss of income from crop cultivation. The income from Bajra cultivation is estimated at about Rs. 600 per *bigha* while the income from grass is estimated at Rs. 250 per *bigha*. The cost of *Adav* is, therefore, about Rs. 350 per *bigha*. This is to be seen against the higher income from Bajra in the subsequent years due to regeneration of soil fertility.

ANALYSIS

Survey

Technique used	*Number of respondents using it*	*Reasons*
Traditional: *Adav*	8	Higher crop yield in subsequent years. Erosion of soil is prevented. In winters cattle eat the grass growing there.
Modern: No *Adav*	2	Cultivate own field and take *Adav* on rent from others for four months of rains.

It will be seen that even those who are not leaving *Adav* themselves are nevertheless using the same technique by taking land on rent from others. Therefore, the use of *Adav* is nearly universal. But the area under *Adav* is reducing. Bal Mukund of Purohitsar used to have 25 livestock and left *Adav* of 60 *bigha*—one-half of his landholding of 120 bigha. Now he has only 10 livestock and has reduced the *Adav* to 40 *bigha*. This decline in number of livestock is in proportion to the reduction in the number of draught animals.

The shift from draught animals to tractors has reduced the need to provide grazing land for cows and bullock. But the farmer does keep other livestock—buffalo, goat and sheep—for milk, meat and cash income. These livestock continue to require grazing land during the rains. Therefore, *Adav* is an integral part of utilization of fodder that is produced along with crops. Crops produce fodder. Livestock are kept to consume the fodder. Grazing land is required during rains for livestock.

The threat to *Adav* comes from sale of fodder. If that takes place then farmers can shift to non-livestock homestead. This possibility, in turn, is linked to the availability of wage employment in the cities. If farmers are able to get sufficient wage employment then they may do away with livestock and undertake absentee agriculture. Livestock generate income and employment as long as the homestead is located in the fields. *Adav* will be threatened when homestead agriculture comes under stress.

Grass plantation on *Adav* does not appear to be an attractive proposition. *Adav* and crop agriculture are mutually dependent on each other. *Adav* is dependent on crop cultivation as the livestock that it serves get fodder from crop stalks. Crop cultivation is, in turn, dependent on *Adav* as it helps regenerate the soil fertility. Therefore, the option of planting grass is not likely to be feasible. Planted grasses like Sevan and Dhaman deplete soil fertility and would prevent crop cultivation. Chunnilal of Dhudsar said that grasses grow spontaneously on croplands recently left as *Adav*. He said that grass seeds come through the wind. Perhaps these indigenous grasses, unlike Sevan and Dhaman, do not compete with crops. On the contrary they provide biomass for *Sood* and *Jhoor* and thereby act as a source of organic manure. Mula Ram of Kinsaria similarly said that *Adav* is left only when other sufficient source of grass such as from common land are not available. He, in fact, took land on rent for providing grazing land to his livestock while he fully cultivated his own croplands.

The purpose of *Adav* is not to provide fodder. It is to enable better utilization of fodder produced during crop cultivation. Trees similarly compete with crop cultivation.

Chunni Lal of Dhudsar was practicing the following rotation:

Year 1	*Tawi*-based crop cultivation.
Year 2	Cultivator-based crop cultivation
Years 3 and 4	*Adav*

It will be seen that *Adav* is integrally linked with crop cultivation. This scheme involves 'double' rest—the topsoil is provided rest by upturning by *tawi*; the whole soil is provided rest during *adav.* It has to be examined whether providing such rest is necessary. Can rest provided by *tawi* alone maintain the soil fertility? The increasing use of *tawi* every three years appears to indicate so. Therefore, *Adav* is likely to decline.

SUGGESTIONS

Undertake study of soil fertility under the use of *tawi*. Examine whether *Adav* is now required.

Fallow land (*adav*) for grazing

Grass harvested from *adav*

This field was left as adav last year

28

Common Grazing Lands (Gochar and Oran)

Category: Potentially Viable

There are two types of common grazing or forestlands in Western Rajasthan—*Gochar* and *Oran*. *Gochar* is the common land of the village, which is so earmarked by the revenue officials. The land was assigned for this purpose by the government—mostly during the rule of the princely states. Any cattle of the village may graze on the *Gochar* land. The administrative control of the land remains with the Panchayat but it cannot be sold or allotted or used for any other purpose. Legally no one can use the *Gochar* land for cultivation or other purpose. It is considered to be public property. There are statutory limitations even on the government's power to allot *Gochar* land for any private or public purpose.

Some elders say that there existed a tradition that cattle were not allowed to graze in the *Gochar* for one or two months during the rainy season so that the grass would grow back. During this period the farmers were to graze their livestock on their own fallow lands—*adav*—left precisely for this purpose. But this rule, if it existed, is now rarely followed because the traditional law-making system has been dismantled and it is not possible for the community to punish the transgressors. The formal judicial system is too expensive, slow and unreliable to provide quick justice in such matters.

Any person creates an *Oran* as a donation to a deity. The *Oran* too is used for the welfare of the people. The ownership, however, rests with the deity to which it is donated and the control is with the presiding priest of the temple. Technically this is private land owned by the deity hence it can be allotted or used for other purposes. The *Oran* is to be used for the welfare of the people but the precise nature of such use can vary from one *Oran* to another depending upon the conventions of the particular temple.

Fear of the deity's curse is an important element in the protection of the *Oran.*

Gochar

Jagirdar Sohan Singh established the *Gochar* of Rampura. He was to lose a large amount of land in land ceiling laws enacted in late fifties. At that time he converted 3,500 *bigha* into *Gochar*. His descendants are living in the village and they 'request' the people not to encroach upon the *Gochar* saying, "Our ancestors have given this land for *Gochar*. Please do not cultivate this land." A fine of Rs. 500 may be imposed on any person who cultivates *Gochar* land. But the revenue officials rarely enforce these laws.

Oran

Folklore has it that the Dahiya king Chachdev founded the city of Kanak Nagar, which is now called Kinsaria. It is he who established the *Oran* of Kevai Mata—a local deity. There was an ancient temple of the Goddess Kevai Mata and the king believed that having an *Oran* would protect the temple. The income from the *Oran* was to be used for the upkeep of the temple and for paying the salary of the priest. Another reason for making the *Oran* was to give protection to wild animals like rabbits, jackals, lions, etc. It is believed that a lion can be seen in the *Oran* during the festival of Navratri. In 2001 a villager Mangala informed others that a lion had come and two other villagers Om Prakash Sevak and Richh Pal Singh saw a lion drinking water in the *Oran* during Navratri. The *Oran* is 1000 years old and has an area of 6000 *bighas*. Earlier it had many trees like Kair, Kankeda, Avail, Board, Dhokada, Rohida, Neem, Khejri, etc. But these are decreasing as the villagers have started to cut trees for burning. In the past those who cut trees were fined. But after independence the people of lower caste started to cut trees to make a living. In the year 1982 under the *Aravali Pariyojana* the government planted Israeli Babool in 1000 *bighas* of land due to which the local trees and grass slowly became extinct. Presently this *Oran* is being destroyed because there are no rules for its management. Previously the priest of Kevai temple had control over the *Oran* land but today no one listens to the priest. There is a lot of cutting of trees especially by lower castes.

The *Oran* at Rampura has only about one-half of the trees left. The rest have been cut. The belief that if one does such a thing his hand and feet will become crooked no longer appears to hold sway. Previously the village headman used to impose fine on those who would cut trees. One Basin Khan had cut one tree about ten years ago. Mangal Singh, the headman at that time, had imposed a fine of 12 kg of Bajra to be fed to the birds. But now the headmen do not wish to take this responsibility upon them because their action can be challenged in the courts or before the revenue authorities.

When are they used?

The *Gochar* and *Oran* are used throughout the year. A meeting of village elders at the beginning of the rains would 'close' the *Gochar* for grazing for a limited period to enable the grass to grow again.

Where are they established?

Gochar and *Oran* are established in the village lands. *Gochar* is revenue land earmarked for the purpose. *Oran* is private land donated to a deity.

How are they established?

The government or the donor establishes the *Gochar* or *Oran*. The Panchayat or the priest of the temple makes rules for its use. A guard could be appointed for ensuring that no cattle enter the *Gochar* or *Oran* during the prohibited period. In some villages every day one person from a family is assigned the duty of guarding the land instead of appointing a guard. It was considered to be a social obligation for each family to offer its services. Previously, the ponds in the *Gochar* too were cleaned periodically. Villagers would do this task together as a community service.

Cost

No cost is involved, as *Gochar* and *Oran* mainly require social management. Traditionally tree or grass plantation is not done and natural regeneration is encouraged.

Effect

Cattle do not have to go long distances in search of grass. People get fruit like Ber, Kumatiya and Sangri free from the *Gochar* and *Oran*.

Trees bring rains. Rain leads to growth of grass, which is then eaten by cattle.

Many *Oran* have *Nadis*, which provide water to cattle as well as people.

Festival

The meeting of village headmen is fixed generally on the day of *Amavasya* or *Poornima*. *Amavasya* is preferred as people do not work on this day and they are free to attend the meetings.

In most, if no all, *Oran*, a fair is held on some specified day to match the festival of the deity concerned. At Kolu Paabuji on the tenth day of Bhadva a fair dedicated to Paabuji is organized in which people of all religions participate. A *Jaagran* (night spent in praying) for the deity Paabuji is organized by the Bheel tribe. Handmade pictures of Paabuji called *Paabuji Ki Pad* are placed behind a curtain. Then the Bheel religious leader—called Bhopa—explains the significance of the pictures to the devotees with the help of interesting tales.

At the *Oran* in village Ugras women enjoy themselves on swings in the month of *Shravan*. On the day of Diwali the fruit brought from *Oran* is worshipped. It is believed that if this were done then there would be no disease in the village for a year. Wood is brought from *Oran* to build the Holi fire.

ANALYSIS

Almost all the people use the *Gochar* and *Oran*. A survey found that only one person out of twenty said that he was not using these common lands.

Survey

	Number of respondents using/not using	*Reasons*
Traditional : *Using Gohar and Oran for grazing*	19	(1) Cattle get grass to eat. In summers cattle get leaves of trees to eat. (2) Cattle can wander freely without fear of being lost. (3) Cattle get shade to sit in. (4) Fruit is obtained from *Oran* for humans and cattle. (5) Cattle can get drinking water from Nadis.
Modern : Not using *Gochar and Oran* for grazing	1	Too many cattle in *Oran* can lead to disease.

Cow and buffalo

Cows were required in the traditional agricultural system, in part, to beget bullocks for agricultural cultivation. The role of the bullock has receded with the introduction of tractors. The farmers find it more convenient to keep buffaloes for their requirement of milk since they do not require to be taken for grazing and the milk also has higher fat content. Cows require to be grazed while buffaloes do not. The very need of a *Gochar* or *Oran*—to make available grazing land—has, therefore, been removed. The farmer now has little need for the community grazing land. He is more interested in encroaching upon it and converting it into a private cultivable land.

From the collective viewpoint the conversion of common grazing land to private cultivable land may not be detrimental. The total biomass produced and fodder made available from cultivation of Bajra appears to be as much, if not greater, than grass. Therefore, conversion into cultivable land is more beneficial to the community as a whole.

The farmers are moving towards keeping a larger number of goats and sheep instead of cows and bullocks in order to make the best use of the available grazing land. Goats are known to engage in hard grazing. The result is that the growth of trees, shrubs and grasses has suffered. The farmer has no incentive to protect the grazing land and maintain it in good condition for the cows and bullocks.

The three alternative uses can be ranked as follows from the standpoint of availability of fodder from the common grazing land:

High	Agricultural or other cultivation
Medium	Grazing by cows and bullocks
Low	Grazing by sheep and goat

Grazing by cows and bullocks is not economically feasible in view of the bullock having been replaced by tractors. Cultivation is not possible due to legal limitations. Thus, the common lands are being used at their lowest returns—by sheep and goat.

Grazing management

The most difficult period for procuring fodder for the livestock is at the beginning of the rains. The agricultural fields are 'closed' for grazing. The common lands are already overgrazed and exhausted through the intense summer heat. The farmer wants to cultivate crops on a maximum amount of his lands. After about six to eight weeks weeds begin to emerge in the fields and *Nidaan* is done. These weeds are used as fodder. This is the first source of fodder after the onset of the rains.

The grass in the common grazing lands, however, begins to emerge within three weeks of the first rains. These are the only source of fodder till the weeds are available. Perhaps it is for this reason that the 'community' does not generally agree to the 'closure' of the grazing land. The opportunity cost of newly sprouting green grass in the first two months is much greater than the grass that may be obtained at the end of the season.

Panchayat and the law

Gochar and *Oran* are community assets. They can only be protected if the community cooperates. The traditional community management system was intermeshed with the system of governance. Both the administrative and judicial powers were vested in the village headman, who normally governed with the consent of the village elders. This arrangement made it possible for the headman to punish those who violated the rules of the community.

The judicial powers have now been transferred to the administrative machinery—separated from the village. The headman has been reduced, at best, to the status of a prosecutor. He has been incapacitated. The power of the *sarpanch* is presently limited to filing a case with the police. He does not have the power to impose punishment in keeping with local traditions.

The elections to the panchayat have had a disastrous effect—at least, in the short run. The headman is not inclined to 'punish' the wrongdoers because it hits at his electoral base. Secondly, the political opponents unnecessarily oppose every good action. The 'community' no longer exists. The entire modern system is based on individual property and rule of law enshrined in the state. There is no role for the community.

This is not to eulogize the 'community'. It is true that this same 'community' did many atrocities against the dalits and the weaker sections. That may have been its negative aspect. But our concern here is to look at the impact of the community on the grazing lands.

Land laws

The distinction between *Gochar* and *Oran* is important. *Gochar* cannot be allotted as it is deemed to be public property but *Oran* can be allotted as it is deemed to be private property belonging to the deity. Part of the *Oran* at Kolu Paabuji, for example, has been allotted for the establishment of market and building houses.

The legal status of an *Oran* is unclear. It is private land for community use. It, therefore, suffers from disadvantages at both ends. It can be allotted as well as it is difficult to persuade the community to observe any rules with regard to its management. This is leading to their degradation. The priests have to fight a lonely battle against the land mafia for no personal gain.

Religious sanction

The decline of *Oran* also owes itself to the decline of faith in the deities in whose name the Oran has been established. Often there is no *pujari* in the temple. The fear of the deity has also become less. The elders believed that cutting trees brought the curse of the deity. Many stories are told about harmful effects of such a curse but there is very little by the way of 'proof'. Perhaps the curse worked because the person violated a socially accepted tradition—or his conscience. If the social tradition is itself finished then there arises no conflict between one's action and one's conscience in cutting of the trees. Moreover, Muslims of the village do not worship the deity and they have less compunction in violating the religious injunction of not cutting the trees.

Cultivation of Oran

The bar on cultivation of *Gochar* is absolute but on the cultivation of *Oran* it is more flexible. The Bhadariya *Oran*, for example, is widely cultivated between the Ber trees. The goats used to eat the Ber fruits. The seed used to get excreted with the *mingani*. The goats used to roam in the fields and spread the seeds. Then the land was cultivated. In this way a large number of trees of Ber have grown. Agricultural cultivation has led to an increase in the number of Ber trees.

Tree plantation

The forest department has planted Babool trees in 100 acres of land and also put up barbed wire in the *Oran* of Kolu Paabuji. Sevan and Dhaman grass have also been planted. The forest department takes care of this part of the *Oran* at present. But this activity does not appear to have the support of the people. The trees planted earlier in the same *Oran* have been an utter failure. Small stubs of trees can be seen in a row (see figs.).

On the other hand, the results have been remarkable at Kinsaria. There is a great difference in the population of trees in the planted and unprotected areas (see figs.).

Tree Plants at Oran of Kinsaria

Serial No.	*Tree species*	*Number of trees in two bigha land (unprotected traditional)*	*Number of trees in two bigha land (protected-planted by Forest Department)*
1.	Neem	1	Nil
2.	Kankeda	15	2 (Plus 4 dead)
3.	Jaal	2	Nil
4.	Kair	30	4
5.	Ber	5	Nil
6.	Israeli Babul (planted)	Nil	35
	Total	53	41

It will be seen from the above table that the tree plantation—which is visibly 'successful'—has actually led to a decline both in the biodiversity of the trees as well as their population.

The central question here is whether such plantation serves the basic purpose of establishing a *Gochar* or *Oran*? Trees compete with grasses unless planted at long spacing of say 100 feet. No grass grows below the Israeli Babul plantation at Kinsaria. Therefore, such plantation may get revenue for the panchayat or the forest department, and it may even be environment friendly, but it negates the very purpose of providing common grazing lands. No successful experiment of plantation of grasses has come to light. At Purohitsar, grass planting had failed.

The approach of tree plantation does not, therefore, appear to hold much potential.

SUGGESTIONS

1. The basic purpose of *Gochar* and *Oran* is to secure public good—in this instance, through making available common grazing land. It has to be, therefore, examined whether grazing land is the best method of securing public good in the present circumstances. It appears that now grazing is a low-priority need in view of the decline of the use of bullocks, which has done away with the need to maintain a large herd of cows. The needs of the people that can be most served by the common land should, therefore, be examined afresh. Monoculture plantations, which provide fuelwood or timber such as Israeli Babul

at Kinsaria, do not provide any immediate relief to the people. In fact, they take away critical grazing land at the onset of monsoons. Alternative people-friendly uses of common lands should be promoted. These could include fruit plantations or water catchment. The suitability of fruit trees of Amla, Kair and Lasoda (Goonda) may be examined in this respect.

2. The legal status of the *Oran* in land laws should be clarified. It may be put at par with *Gochar* to prevent its allotment for private purposes.

3. A social consensus along the lines of Joint Forest Management should be built on the proper maintenance and use of *Oran* and *Gochar*. The relationship of panchayat elections with *Gochar* and *Oran* is tricky. Panchayats are valuable in themselves as a measure of promotion of democracy. But democracy itself works only if there is a non-political agency to define the public good and spread its awareness among the people. This issue should be sorted out.

4. The loss of *Oran* is related to the decline of religious sentiments. A meeting may be organized of the priests of deities to whom *Oran* have been dedicated. Stories of 'curse' of the past and present should be documented and awareness generated. That will help instill fear among the people and restrain them from cutting trees.

Oran at khara

Kair trees at Kinsara oran

Gochar at Kinsara

Neem tree in Kinsaria oran has died after planning of babul trees

Trees planted at Kinsaria Oran by forest department

New plantation at Kalu Bapuji oran

Previous plantation with trees surviving as shrubs

Temple of Kalu Bapuji

Uncultivated part of Bhadriya oran

Uncultivated part of Bhadriya oran

Cultivated part of Bhadriya oran. Tractor has ploughed land between ber trees which are sacred

29

Land on Rent for Grazing (Bhelwad)

Category: Potentially Viable

After the crop has been harvested there is agricultural waste, some grain and grass lying around or growing in the field. This is suitable feed for the cattle. The field is taken on rent by the herdsman to graze their livestock. This is also done for grasslands. Some persons do not have a large number of livestock and they have land that is unfit for crop cultivation. They rent out those lands to the herdsmen.

When is it done?

Fallow land is rented at the beginning of the rains. Cropland is rented after the crop has been harvested.

Where is it done?

Fields are rented for this purpose.

How is it done?

Farmers who have a large area of land and herdsmen who have a large number of livestock meet with each other and agree to a rent.

Effect

The landowner gets some income while the herdsman gets assured supply of feed for his herd for some time. The dung of the cattle, which graze in the field, acts as manure and makes the soil fertile.

Budget

Mangu Singh of Jelu Gagadi rented the following lands in 2001-02:

Plot 1: 150 *bigha* (Bajra was cultivated on 50 *bigha*)	Rs. 2,100
Plot 2: 170 *bigha*	Rs. 2,200

He managed to feed his herd of 250 animals (50 goats and 200 sheep) for 2½ months from November to January on this land. He said that the feed available on cultivated land is about one-half of that available on fallow land.

ANALYSIS

The practice of Bhelwad is likely to continue. There will be some persons who have land but no livestock and vice versa. This trade between the two is mutually beneficial. No alternative to this social technology is available.

Survey

	Number of respondents using the technique	*Reasons*
Using Bhelwad	3	Get assured supply of feed and saved from having to migrate.
These not using the technique	8 (Two persons do not have livestock. Six persons have not ctaken Bhelwad this year because they do not need the same)	Do not have a big herd.

SUGGESTIONS

- It may be possible to develop a more formal land exchange for the benefit of the people. It seems that the land laws do not permit such renting of land. This practice is undertaken informally. A representation can be made to the government to legally allow such renting. The development of a wider market will lead to better utilization of land resources.
- A *Bhelwad Mela* may be organized to make the system more efficient.

Field rented for grazing

XI

Livestock

30

Livestock Nursery (Tutadia)

Category: Stable

Tutadia is a shelter built to protect kids of goats and sheep (and, sometimes, also dogs and cats) from cold and wild animals.

Clay Tutadia

When is it used?

The shelter is built at the onset of the winters when the young ones have to be protected from the cold.

Where is it done?

It is built in the homestead (*Bada*) meant for the sheep and goats or near the home.

How is it done?

Traditionally *Tutadia* can be made underground or on the surface.

Underground

Earth is dug out in the shape of a round cone, which is one to one-and -a-half feet on the top and two feet at the bottom. The mouth of the hole is made firm by placing an old tire or stones on it and smearing it with mud. The kids are put in the pit and the opening is covered with a wooden or stone slab.

Surface

Murad—a type of hard clay—is dug out and made into a mortar. Blocks are made from it. Walls of the shelter are built with these blocks. The height is about two feet and the length and breadth about four feet. Wood strips or sticks of Aak are placed on this wall to make the structure of the roof. Sinia grass is tied on them. Sand is sprinkled over the grass. Finally a paste of dung and clay is made and smeared over the roof. In one of the walls an opening of about 12" is left to serve as the gate. The kids are put in through this gate and then an iron, wood or stone piece is used to block the entrance.

Effect

The *Tutadia* protects the kids from cold as well as rain. They are also safe from dogs and cats in this shelter. With the kids locked inside the shelter, it is easy for the women to milk the goats and sheep. But as the shelter is primarily made of unbaked clay it can crash and cause harm to the kids.

The place where this shelter is built gets the droppings as manure and the soil becomes fertile. If plants are planted there later they grow well.

Generally it is the women who take care of the kids. They do not have to worry for their safety. If the kids remain in the open they tend to wander about and enter into neighbouring houses thus proving to be a nuisance.

Cost

Underground

2 persons for labour for digging @ Rs. 80 per head	Rs. 160
Cover (gate) for the opening	Rs. 50
Cement or tire	Rs. 100
Total	**Rs. 310**

Surface

2 persons for labour for making blocks and walls @ Rs 150 per head	Rs. 300
1 person for labour for cutting Sinia	Rs. 80
Cost of wood	Rs. 100
Cost of putting a layer of dung and clay on roof	Rs. 100
Cost of *Murad* or hard clay	Rs. 400
Total	**Rs. 980**

Festival

Earthen lamps are lit in the shelter on the night of Diwali.

STONE TUTADIA

How is it done?

The walls are first made of stone. Stone slabs or strips are put over the walls to make the roof. The whole structure is smeared with clay or dung. A small gate is left in one wall as entrance. The kids are put inside and the opening is blocked with a stone slab. When the kids need to be taken out the slab is removed. The size of the shelter is such that the kids can comfortably sit inside.

Effect

This *Tutadia* lasts several years. There is no danger of it collapsing in rains. Once the shelter is built the family members do not have to worry about the safety of the kids and they do not have to keep a constant vigil. But this structure does not provide much protection from heat or cold as the stone slabs get hot or cold.

Cost

Cost of stone	Rs. 250
Cost of labour	Rs. 200
Total	**Rs. 450**

GRASS TUTADIA

How is it done?

This shelter can be round or rectangular. Around the demarcated area the earth is dug out and wooden poles or sticks of Aak or other plants are erected to make the walls. The sticks are tied with ropes to make the structure stable. Pai and Sinia are used to make the roof. On one side a gate is left to put in and remove the kids.

Effect

This structure is not as safe as those of stone or clay. It also does not provide as much protection from heat and cold as its walls are porous.

Cost

Cost of wood	Rs. 50
Cost of Sinia and Pai	Rs. 50
Labour	Rs. 80
Total	**Rs. 180**

STONE AND CEMENT TUTADIA

At present permanent shelters of stone and cement are built. These last many years and are stronger.

How is it done?

Permanent shelters are made from stones and cement.

Effect

Such a shelter is strong and does not collapse in heavy rain. It keeps the kids absolutely safe from cold, rains, and animals like dogs and cats, which can kill them. But they do not provide protection against heat and cold.

Cost	
Cost of one trolley stone	Rs. 800
Cost of 5 bags of cement @ Rs. 150 per bag	Rs. 750
Cost of two persons for labour @ Rs. 150 per head	Rs. 300
Cost of stone strips	Rs. 300
Total	**Rs. 2150**

IRON COVER AS TUTADIA

An umbrella-like cover of iron (see fig.), often used for keeping hens, is placed over the small kids.

How is it done?

An iron cover made of strips is placed above the small kids when they have to be protected, especially during the night.

Cost

The cost of an iron cover is about Rs. 100.

Effect

The iron cover can be used only for very small kids. Its utility is limited.

NO SHELTER

At present many people do not build any shelters. The kids are allowed to roam freely generally in the house itself or out in the open.

Effect

The kids do not remain safe. They catch cold easily in winters and get ill. They also do not leave much milk for the family as they suckle most of it from the udders of their mothers. Women have to worry a lot about them and have to leave their chores to take care of the young ones.

ANALYSIS

Survey

Type of Shelter / No Shelter	*Number of respondents*	*Cost*	*Reasons*
Clay (Underground or on Surface	8	Underground Rs. 310; on surface Rs. 310	Good protection from cold
Stone	9	Rs. 450	Less protection but easy to handle.
Grass	2	Rs. 180	Low protection, low cost but high maintenance.
Stone and cement	2	Rs. 2,950 that	Strong shelter protects kids from cold, rains and other animals
Iron	NA	Rs. 100	Suitable for very small kids only
No shelter	1	Nil	

The clay or stone *Tutadia* are used by a large number of farmers. They provide good protection and are durable. The modern technology of stone and cement does not provide protection from heat and cold. This technology appears to be quite stable.

SUGGESTIONS

A technical survey of clay and stone *Tutadia* may be made. Design improvements may be suggested.

Underground *tutadia* with tire as mouth

Underground *tutadia* with stone mouth

Surface earthen *tutadia*

Grass *tutadia*

Iron cover

31
Shelter for Livestock (Taap)

Category: Potentially Viable

Shelter is built to protect livestock from cold in winters and the heat and sun in summers. The shelters also protect them from wild animals.

TAAP

Taap is a shelter made of grass and bushes.

When is it built ?

It is built just before the start of the winter season. Or it is also built before summers to protect livestock from the hot wind called *loo*.

Where is it done?

It is made near the home.

How is it done?

An area is demarcated for the *Taap* according to the number of livestock. Then earth is dug along the boundary of this area to erect poles of wood, Aak or other sticks. Rope of Sinia is used to tie the sticks with one another to form a fence-like enclosure. Another ring of sticks of Aak are tied to make a wall. The Aak sticks are slowly sloped inward to make a dome-like structure. *Pai* (sticks of Ber) may be used in the roof area.

The shelter is higher on the north side and lower on the south. Door is on the south side so that livestock do not feel cold.

Effect

The livestock are protected from cold, rain and hot winds (*loo*) and stray animals. The dung of the livestock can be collected and used as manure. The family members remain free of worry with a protective shelter around the livestock. It is comfortable for women to milk the livestock in the shelter, as it is warm in the winters and cold in summers.

Cost

The cost of a typical *Taap* suitable for five or six cattle is estimated as follows:

Cost of 50 wooden poles @ Rs. 5 per pole	Rs. 250
Cost of cutting and transporting Sinia	Rs. 300
Cost of Aak sticks	Rs. 200
Cost of five persons for labour for cutting Aak @ Rs. 60 per head	Rs. 300
Cost of Pai	Rs. 240
Ropes of Sinia	Rs. 400
Cost of eight persons for labour for making Taap @ Rs. 70 per head	Rs. 560
Total	**Rs. 2250**

Festival

On the day of Diwali *Govardhan Pooja* is done in the *Taap*.

Dug-In Taap

The floor of the *Taap* is dug out to form a low-lying area; or a sand dune is used as part of a wall. This protects the livestock better from cold. The livestock are more comfortable and warm here.

Where is it made?

It is made some distance from home in a place where there are bushes or a *Tibba* (dune) that acts as a wall on one side.

How is it done?

Earth is dug out from the demarcated area to make a pit at least three feet deep, 15 feet long and 10 feet wide. Rest of the procedure is the same as in the case of the normal *Taap*.

Effect

The livestock have better protection from cold and heat.

Cost

Cost of digging: one person for labour for 3 days @ Rs.80 perday	Rs. 240
Cost of 50 wooden poles @ Rs. 3 per pole	Rs. 150
Cost of cutting and transporting Sinia	Rs. 150
Cost of Aak sticks	Rs. 150
Cost of labour for cutting Aak	Rs. 200
Cost of *Pai*	Rs. 200
Ropes of Sinia	Rs. 300
Cost of labour for making *Taap*	Rs. 480
Total	**Rs. 1870**

The cost of a dug-in Taap is lower than the normal Taap because the built structure is smaller. But this can be built only if a sand dune-like structure is available near the homestead. Probably digging is not done on level surface because it may lead to flooding in the rains.

JHOPA

A *Jhopa* or hut is made to protect the livestock from cold in winters and heat in the summers.

How is it done?

The *Jhopa* can be rectangular or round in shape. Wooden poles are erected in the ground around the area chosen for the shelter. Sticks of Aak or other plants are tied to form a strong wall. A roof of Moonja supported by wooden *Patti* (strips) or other long sticks is made. The roof of *Taap* is made of Aak sticks in one continuous structure of wall and roof. The roof of a *Jhopa* is made of grasses while its walls are made of Aak sticks.

Effect

The roof of Jhopa provides better protection from rains as well as cold. Women too help in the building of the *Jhopa.*

Cost

Cost of wooden poles	Rs. 300
Cost of Moonja	Rs. 500
Cost of sticks	Rs. 100
Cost of 4 persons for labour	Rs. 300
Cost of Patti	Rs. 200
Total	**Rs. 1400**

The cost of erecting a *Jhopa* is less but its maintenance requirement is greater. The roof has to be rebuilt or reinforced every year.

CHHAPRA

A simple shade is built like a canopy extending from a wall. This shelter has no sidewalls on three sides.

Where is it done?

It is made in the backyard of the house or adjoining any wall of the house.

How is it done?

This shelter is built with a wall of the house acting as support. Wooden poles about 10 to 12 feet high are erected in the ground about 15 feet from the wall. Then from the wall to these poles long pieces of wood are tied to form a skeleton for the roof. Small sticks and wood are placed is this skeleton and tied. Then on the top go dried bushes, grass and Aak leaves to form a roof. These are placed properly and tied so that there is no space left in-between.

Effect

It is easier to feed the livestock and water them in this shelter. But the protection from cold and wind is relatively less.

Cost

Cost of building such a shelter of 25 feet x 15 feet x 10 feet is about Rs 800.

TAAP WITH STONE WALLS

One modern variation is to build the Taap with walls of stone strips. Such a structure requires less maintenance and lasts longer.

How is it done?

Stone strips (about 12 inches wide and 6 feet long) are erected along the boundary of the proposed shelter. A roof of Sinia and Kheep is built supported by strips of wood. To tie the roof plastic ropes are used instead of ropes made from Moonja.

Effect

These shelters are stronger and last longer. However, they have a definite disadvantage for the livestock. The walls of stone remain cold in winters and hot in summers.

Cost

Cost of 7 stone blocks	Rs. 500
Cost of 4 persons for labour @ Rs. 50 per head	Rs. 200
Cost of wood sticks	Rs. 200
Cost of transportation	Rs. 500
Cost of Kheep, Sinia	Rs. 500
Total	**Rs. 1900**

TAAP WITH TIN ROOF

These shelters are built with walls of stone and roofs of corrugated tin sheets. Some people even have walls built of bricks and cement.

How is it done?

First stone blocks are made to stand erect along the boundary of the proposed shelter. A wall of stone is obtained. Wood is also put around to fortify the wall. Generally it is done on three sides with an outside wall of one's home acting as the fourth wall. Over it corrugated iron or tin is put as roof. Then the shelter is ready.

Effect

These shelters are stronger and last longer—up to ten years, while that with grass roof has to be rebuilt or repaired every year. Shelters made of wood and grass can also be destroyed by fire while there is no such danger for a shelter with walls of stone and iron roof. However such shelters remain cold in winters and hot in summers—even more than those with stone walls alone. Besides, building such a shelter involves more cost and labour.

Cost

Cost of 4 stone blocks	Rs. 600
Cost of 2 persons for labour @ Rs 60 per head	Rs. 120
Cost of wood	Rs. 200
Cost of 7 iron sheets	Rs. 1750
Cost of building	Rs. 200
Total	**Rs. 2870**

WALL AS SHELTER

Many people today do not build any shelters; rather they keep the livestock tethered near some wall.

Where are the Livestock Kept?

Livestock are tied on the side of the house which is most protected from cold winds in the night.

How are the Livestock Tied?

Livestock are tied on the leeward side of the house so that they are protected from cold winds in winter.

Effect

The livestock are not properly protected from cold in winter and they suffer from cold and even pneumonia. As a result the family members have to worry and take extra care of the livestock.

Cost

No cost is involved.

ANALYSIS

The comparative usage, cost and merits of the various techniques are given below :

Type of Structure	*Number of Respondents using the structure*	*Cost*	*Analysis*
Taap	15	Rs. 2250	High cost of maintenance. Leaks in rains.
Dug-in *Taap*	5	Rs. 1870	Availability of suitable sand dune limits its use.
Jhopa	4	Rs. 1400	High cost of maintenance. Suitable for the poor as it is made almost wholly by the use of own labour.
Chhapra	9	Rs. 800	Less protection from cold and heat. No cash expense is involved. Suitable for persons having a few livestock.

Contd.

Type of Structure	*Number of Respondents using the structure*	*Cost*	*Analysis*
Taap with walls of stone	7	Rs. 1900	Less maintenance. Less protection from heat and cold.
Taap with tin roof	5	Rs. 2870	High cost. Low maintenance. Lesser protection from heat and cold. No danger of fire.
Wall as shelter	4	Nil	Low cost. Least protection from heat and cold. Probably involves higher medical expenditures.

It is clear that traditional technologies continue to remain dominant. The traditional technologies are labour intensive. They require intensive repairs every year. But they involve less cash cost. Stone or tin sheets do not have to be purchased. The main factor influencing their use is low wage. On the other hand, modern technologies are less effective in providing protection from heat or cold. But they require less maintenance though incurring high cash cost. It may be presumed that the medical expenses for livestock are greater in the modern technologies. It seems that those who have cash incomes coming from the city and whose opportunity cost of labour is higher prefer modern technologies.

SUGGESTIONS

Try combination of modern and traditional technologies. The main problem with modern technologies is that they provide less protection from cold and heat due to stonewalls and tin roof getting cold and hot. This could probably be managed by placing Moonja or other grasses on the top of the tin sheet. A survey of construction material may be made to find material which provides greater insulation than stone slabs. The objective should be to provide relief to the farmers from the yearly labour of maintenance while also reducing cost and ensuring durability of the structure and its effectiveness in providing protection.

Taap

Roof of *taap*

Taap under cons-truction

Dug-in taap

Jhopa

Chhapra

Taap with tin roof

Using wall as shelter

Using wall as shelter

32

Fodder Shredding (Kutar)

Category: Stable

The stalks to be used as fodder are shredded before feeding them to the livestock, as they cannot eat whole plants. This shredding is called *Kutar Katana.*

When is it done?

Sometimes farmers shred the stalks before storage. At other times they store the whole stalks and shred as much as needed for feeding to the livestock on daily basis.

Where is it done?

It is done in the homestead (*Bada*).

How is it done?

The stalk is kept on a wooden platform called *Nigah.* They are cut into small pieces with a hand tool with an iron blade and wooden handle called *Jharbat.*

Cost

One person can shred five bunches (*Oda*) weighing about 25 kg each in one day. The wage rate is taken at Rs. 60 per man-day. The cost of shredding (manually) comes to 48 paise per kg.

Effect

Livestock get fodder in the form of small pieces that are easy to eat. They eat the hard part of the stalks in this manner, which they otherwise leave.

SHREDDING BY TRACTORS

A shredder is powered by the tractor and used to shred the fodder. It saves time and energy. The hard parts of the stalk are thrown away when shredding is done by hand because it takes much greater strength to cut them. But the tractor shredder, a modern machine, turns all parts of the plant into small bits. Thus no part of the plant is wasted.

Women benefit from tractor shredding as they are excused the labour of cutting the fodder with hand or of bringing grass from outside for the cattle.

One disadvantage of the tractor shredder is that it pulls in the plants very fast and there is a chance of mishap as the hand of the worker could be pulled in and cut.

Cost

The hourly rent for a tractor shredder is about Rs. 250. About 60 bunches (*Oda*) weighing 25 kg each can be shredded in one hour.

Rent of Tractor shredder for 1 hour @ Rs. 250	Rs. 250
Cost of 5 persons for labour for 1 hour @ Rs. 20 per hour	Rs. 100
Total	**Rs. 350**
Cost per kilogram (Rs. 350 for 60 *Oda* x 25 kg = 1500 kg)	23 paise

ANALYSIS

The cost of *Kutar* is about one-half by machine in comparison to the traditional manual method. But if the opportunity cost of labour is about one-half of the going rate then the two techniques are comparable in cost. The survival of this manual technique depends upon the wage rate remaining low. However, the technique of *Kutar* is stable. Fodder will continue to be shredded before feeding to the livestock.

Survey

	Number of respondents doing it	*Reasons*
Kutar Katana with Jharbat	Nil	Very tiring and expensive.
Kutar Katana with Tractor Shredder	20	Women are saved from hard labour of doing *Kutar*. All parts of the plant are cut and no part is wasted.

SUGGESTIONS

Improved design of mechanized shredder may be made. A lighter

Jharbat may help increase productivity of the traditional *Kutar*. But the odds are heavily arrayed against undertaking manual *Kutar* by *Jharbat* because it involves more hard work and is expensive. On the other hand, the tractor shredder cuts even hard parts of the stalk and is, therefore, both cheaper and efficient.

Kutar in progress

33

Open Fodder Storage (Kalar)

Category: Stable

The fodder has to be stored safely to prevent it from getting wet in rains. So the fodder is piled together in a way that rainwater does not do much damage. This is called putting up a *Kalar*.

When is it done?

It is done after harvest of *kharif* crop—mainly Bajra—in November and December.

Where is it done?

It is done near the homestead wherefrom it is easy to take and feed the livestock yet it is safe from stray animals. The place where the shelter is built must be level so that it does not topple. The earth is hardened so that termites do not attack the *Kalar.*

How is it done?

The dried fodder is tied in bundles and piled together. Then the bundles are stacked one upon the other in a tight manner so that water does not penetrate. The top is given a slope so that the rainwater slides away and does not enter the Kalar.

Some farmers simply pile up the fodder without arranging them properly in a *Kalar*. This is generally done when they expect the fodder to be used much before the onset of rains. The *Kalar* is made in November-December while the rains are expected in June-July—about seven months later.

Effect

Fodder is made available all the year round. It is protected from rains as well as termites.

ANALYSIS

There is really no alternative to making a *Kalar*. It would be possible to store the fodder in a huge hall but that would be prohibitively expensive. A survey showed that most people having a large amount of fodder did make a *Kalar*. Those not making a *Kalar* said that there was not enough fodder to take the trouble of building a *Kalar*.

Survey

	Number of respondents using/ not using Kalar	Reasons
Making a Kalar	16	Fodder is protected from stray animals and cattle. It is protected from termites. Fodder does not get spoilt for a long time.
Just piling together and covering with branches or thorns	10	Building a *Kalar* takes a lot of time and effort. There is not enough fodder to take the trouble of building *Kalar*. It is easy to take fodder out of a pile than from a *Kalar*.

The technique of *Kalar* is stable and there is no threat to the same. One possible threat would emerge when farmers might have the stalks shredded upon harvest and then store the shredded fodder in covered godowns. That would save much manual labour—both in shredding as well as in stacking and removing. At present the cost of covered storeroom appears to be quite high to make this practicable. But this may change with increasing levels of income and higher wage rates. On the other hand farmers report that the shredded fodder gets infected by a disease called *jala* after five or six months. Therefore, *Kalar* is likely to remain in existence.

SUGGESTIONS

An in-depth survey of various *Kalar* may be undertaken to study the designs and processes. An Institute like the National Institute of Design could be asked to suggest improvements in design of the *Kalar*.

Tying *dhoka*

Throwing bundle to the top

Half-made *kalar*

Completed *kalar*

34
Storage of Dung Cakes (Thepadi)

Category: Declining

Dung of cows or buffaloes is shaped into thin cakes and these are dried. These are called *Thepadi*. The *Thepadi* are stored to be used as fuel during the rains.

When is it made?

Thepadi is made during the winters and it is stored during the summer, before the onset of monsoons.

Where is it done?

Thepadi is stored in the courtyard or in the fields near the homestead—not very far from either the livestock shelter or the kitchen.

How is it done?

Dung of cow and buffalo is taken and shaped into round cakes, which are then allowed to dry. These are collected together and stacked like a small mountain. The stack may be built on a stone slab, which is placed on small pieces of stone. This is done to prevent ingress of moisture. The stack is covered with a thick smearing of cow dung to form a large covered mound.

Cost

The only cost involved is that of own labour. If it were done through outside labour then Rs. 100 would have to be paid as wages.

Effect

Women do not have to wander here and there in search of fuelwood.

On the other hand, the dung, which is used to make *Thepadi*, could have been used as organic manure in fields.

ANALYSIS

As far as the technology of storage is concerned, farmers have already started to create a synthesis of traditional and modern technologies. They have used polythene sheets as a supplement to dung covering.

The more basic question, however, is that of use of *Thepadi* as fuel. The modern option available is to use LPG gas or kerosene stove for cooking instead of *Thepadi* as fuel. The modern techniques appear to have the following advantages:

- There is less labour for women.
- Stoves produce little smoke in comparison with *Thepadi*.

On the other hand, the disadvantages are as follows:

- Kerosene and LPG are not always available.
- Cash expenditure is required in purchasing kerosene or LPG.
- There are limits to the use of fossil fuels.

The traditional technology appears to be both unfavourably poised toward women as well as the soil.

The issue is interrelated with the reliance on fossil fuels for our energy requirements. In the immediate context, fossil fuels are available and it is more important to use dung for manure than as fuel. But there are limits to how much reliance should be made on fossil fuels.

Thepadi is certainly hard work for women. A stove is much more comfortable and, emitting less smoke, also health-friendly.

SUGGESTIONS

This traditional technology should be discouraged. It belongs to a period when other environment-friendly sources of fuel were not available.

Thepadi stack

35

Migration of Livestock (Palayan)

Category: Declining

When there is no feed or grass or plants available for the cattle people leave their villages with the animals to places where there is greenery or feed available. This is called Migration—or 'escape'.

When is it done?

Migration generally takes place when there is no rain and the crops fail. It may also take place when there is less rain and the farmer sows the seed but then there is no further rain and the farmer is doubly hit—he loses the cost incurred on sowing and has no fodder growing for feeding the cattle. Farmers leave in desperation to find work in brick kilns and those with cattle leave for far away places—typically for South Rajasthan, or the states of Haryana and Punjab where irrigation is more widespread.

Migration can also be a regular annual feature in some parts. There are parts of Rajasthan like Jaisalmer where cattle rearing is the main occupation. Grass is available for grazing only during the rains. Thereafter livestock are taken to other parts of the country.

Migration may start as early as November after the crops have been harvested. But the stubs growing in the fields—now open for grazing—keep the animals alive for some time. It is more frequently undertaken during the months of March to July. The farmers of the host areas have irrigated lands. They cultivate the *rabi* crop and the land remains empty in March-April. Livestock can graze the stubs left over after harvesting. This is the most difficult time for the livestock in Rajasthan.

Where do farmers go?

Farmers leave their homes for other areas. Previously they would leave for Sindh and Multan. Now they leave for Haryana and Punjab.

How is it done?

Before undertaking migration an 'advance' person would be sent to the possible destination in order to confirm whether or not there was grass there. He would also choose a good path so that the cattle get something to eat even on the way to the place. In the past it took them two months to reach their destination. They would travel on foot and the cattle would eat whatever they found on the way. In the day they would travel and at night the cattle would be allowed to graze while the men would make camp. If they came across a pasture in daytime they would stop and allow the cattle to graze.

Modern Technology

Two or three farmers jointly rent a truck and load their livestock and leave for places where there is food, water and also jobs.

At least two or three people go with the cattle herd. While one minds the cattle others do labour and earn money and food for all of them.

Cost

The cost of migration is entirely unpredictable. It depends upon whether the herdsman gets payment for sitting his herd en-route or he has to pay cash to buy feed for his livestock.

The cash cost of migration by truck is certainly higher but a comparison is not possible because foot migration can be profitable too.

Festival

Sugan is seen and an auspicious day and time chosen for departing.

Effect

A survey at Baap showed that all herdsmen were migrating.

Survey

Migration Undertaken	*Number of respondents involved*	*Reasons*
Traditional: Migration undertaken	8	To get food for livestock.
Modern: Migration not undertaken	2	Do not have any cattle.

Herdsmen at Kyasaria near Shergarh, however, said that only one-half of the herdsmen migrated. Generally migration was not liked but rather it was a compulsion. Those having tubewells are able to grow wheat and that provides fodder for livestock during the summer months.

- Farmers in the host areas pay in cash or kind to the herdsmen for 'sitting' the sheep in his lands. About one quintal wheat may be given for sitting 200 sheep for one week. The sheep graze during the day and sit on his field during the night. The droppings fertilize the soil. Some herdsmen estimate that if sheep eat and sit in a field for 15 days the yield increases by one-half.

- Sheep and goats leave nothing for cows or other cattle. There is a saying to this effect in Rajasthan.

Oont Chhote Ankada, Bhed Bakri Chhode Kankada

(The camel leaves the Aak shrub. Sheep and goats leave only stones.)

- Males of the family are away with the cattle for long periods. If women remain behind in their homes they have to take care of the children and the entire burden of the family is on their shoulders. They have to arrange for food and water without any male to help them. They also feel lonely and have to face fear and uncertainty. They also keep worrying about their absent husbands. There are several songs which these forlorn women sing remembering their husbands and which depict the state of mind of the departing menfolk as well.

Ham Chale Pardesh Gouri Yad Rakhna,
Kamare Me Beitthi Beitthi Kagad Likhana.
Sasuji Ave to Thoda Khyal Rakhana,
Suseroji Ave To Thoda Man Rakhana.
Ham Chale Pardesh...
(O Sweetheart I leave for foreign places remember me,
Sitting in your room write me letters,
If your mother-in-law comes take care of her,
If your father-in-law comes give him respect,
I leave for foreign places...)

Another song goes like this:

Ud Ud Re Mhara Kala Re Kagla,
Kad Mharo Peevjee Ghar Ave.
Sone Roope Se Thadi Chonch Madaoo,

Pagalya Me Payal Ghadaoo Kaga.
Jad Mhara Peevji Ghar Ave.
Kheer Khand Ra Jeemann Jeemaoo,
Katore Me Doodh Peelaoo Kaga,
Jad Mhara Peevji Ghar Ave.
(Fly O black crow my friend,
And bring me news when my husband would be home.
If you do so I will get your beak plated with gold,
And give you anklets to wear in your feet,
When my husband is back home.
I will give you Kheer and sweets to eat,
And milk in a bowl to drink,
When my husband is back home.)

Another goes like this:

Have Re Deewana Mhane Ekladi Mat Chhodi Re Deewana Tharodi Marwad Me
Pardes Ja ke Pardesiya Bhool Na Jana Piya...
Pardesi Pardesi Jana Nahi Hame Chhod Ke...
(O my lover do not leave alone…
O my lover do not forget me when you are away to foreign lands…
O wanderer to foreign lands do not leave me alone…)

The last two songs indicate that film songs are replacing folk songs these days.

ANALYSIS

Migration is undertaken to manage the fodder requirement during the dry period. Fodder is, however, available during the rains in the home area. The decrease in the availability of grazing land indicates that this proposition may no longer hold. The availability of wage employment is also reducing the willingness of herdsmen to undertake the difficult task of migratory animal husbandry.

The role of animal husbandry in the rural economy is also reducing. Previously animal husbandry was the *major* occupation of the people of the area and agriculture was incidental. The virtually unlimited lands were free for him to graze who could manage a large herd. The limiting factor was management, not availability of grazing land. Now, wage employment and agriculture have become the main occupation and animal husbandry has become incidental. Therefore, it seems that migration would decline in the times to come.

SUGGESTIONS

The key question is how to use the biomass of the area. There is a major shift from cows and bullocks to (1) buffalo (for the use of crop fodder) and (2) sheep-goat (for the use of grazing). Of these, only sheep and goat are made to migrate. The number of these small animals would be reducing in proportion to the decline of common grazing lands. Therefore, this practice is likely to reduce in the times to come.

Sheep migrating

XII

Culture and Festivals

36

Jawari

Category: Potentially Viable

Jawari is an omen. It is done to find if the coming crop would be good or bad.

When is it done?

It is done before sowing a crop of wheat or barley during *Navratri* in October. It is done on *Amavasya, Dooj* or *Teej* or before leaving for sowing seed in the field. It is also done during the festival of *Gangour* in March.

Where is it done?

Farmers do *Jawari* near the *Pareenda* or the place where water pots in the house are kept. It may also be done near a well.

How is it done?

A handful of seeds of wheat or barley are sown and watered. This is done in the earth near *Pareenda*. Alternatively some soil may be put in a shallow earthen pot and seeds sown in it. The growth of the seed is taken to indicate how the crop would be in the future.

Effect

The belief is that *Jawari* indicated whether the crop would be good or not. But there are many other benefits. Jawari indicated whether the seed might have been spoilt due to moisture or insects. It was also a way of learning to save some of one's own seed if only for the *Jawari*. *Jawari* is done with old and not new seeds. Thus it is linked more with getting a pre-assessment of germination. Notably, *Jawari* is done in October before the sowing for *rabi* is undertaken. The purpose of *Jawari* during *Gangour* is not clear.

Sometimes mixed Jawari is undertaken. Seeds of wheat and barley are mixed and sown together. This might indicate which of the two crops may give better results in the current season.

Cost

About 100 grams of seeds are used.

Festival

Jawari is done during the festivals of *Navratri* and *Gangour.*

ANALYSIS

The relationship of *Jawari* with agriculture is not clear. It seems that the knowledge has been lost. A survey shows that only one out of six farmers is undertaking *Jawari* at present.

Survey

	Number of respondents doing/ not doing Jawari	*Reasons*
Traditional: Jawari	1	*Jawari* is done for worship of *Gangour.* It gives prior indication about the future of the crop. *Jawari* indicates the amount of water needed by the crop and whether the seed suits the soil or not.
Modern: *Jawari* not done	5	Used to do *Jawari* earlier. Now they buy seed directly from market, which they presume to be good.

The fact that farmers said that they have given up doing *Jawari* because they now buy seed from the market confirms the utility of the process as a preassessment of germination.

SUGGESTIONS

- Start the practice of doing *Jawari* with seed that are purchased or otherwise brought from outside.
- Educate farmers that *Jawari* is a method of pre-testing seed. Encourage

undertaking of *Jawari* with various seeds. Only wheat and *Jawar* seeds are used now. Instead *Jawari* may be undertaken with seeds of local crops.

- Undertake Jawari in March *Navratri* with seeds of *kharif* crop and in October *Navratri* with seeds of *rabi* crop. Discuss this with religious leaders and disseminate the information among people.
- Start practice of doing *Jawari* during *Gangour* with mixed seeds of Bajra, Guwar, Moong and other crops. This will help assess the quality of seed for the *kharif* season.
- Discuss with scientists whether more useful information can be obtained from the colour, growth and other features of the small plants.

Jawari

Jawari

37

Meteorological Predictions (Sugan)

Category: Potentially Viable

Dry land farming is highly dependent on rains. If rains are good then Bajra gives better profits. If rains are poor then coarse crops like Guar or Til are more profitable. Success in agriculture depends upon being able to predict the nature of rains. It appears that this prediction was done through oracle sayers or *Sugani* who could understand the good and bad omens. Originally the purpose may have been to be able to predict the nature of forthcoming rains and plan crops accordingly. At present, however, this predictive and productive function of *Sugani* appears to have been lost. It has now become merely a method of predicting whether the crop will be good or bad without any implications for crop planning.

When is it done?

Sugan can be seen on the following days

- *Amavasya* of *Vaishakh*
- Second day (*Dooj*) of *Vaishakh*
- On the day of *Akha Teej*
- *Navami* of *Ashad*
- At the first rain
- When beginning cultivation by ploughing.

Where is it seen?

Sugan is done at home when leaving for the fields for the first time in the season or in the field itself before entering it with the plough. Often a Sugani will read the omens at his own convenience such as when going to the field in the morning.

How is it done?

Many methods are used to see *Sugan*. These are specific to the *Sugan*. One *Sugan* may look at the birds while another may look at the bullocks.

- It is said *Khar Daavaa Vish Jeevanaa,* i.e. , when going to the field if a donkey is seen on the left side and a snake on the right it is believed to be auspicious. If a donkey brays in the front it is not considered to be a good omen. But if it does so in the back it is considered auspicious.

- *Sugan* is seen through the bird called *Sugan Chidi,* which has white black stripes. If it chirps sitting on a green tree it is auspicious. If it chirps loudly while flying it is considered inauspicious.

- *Teetar* is also used to see *Sugan*. If a *Teetar* makes a sweet sound in the greenery or at sunrise it is considered good.

- *Sugan* can be known from bulls too. If a bull in the field urinates and walks forward it is auspicious. If it walks forward the first time it enters the field it is good; if it turns back it is bad.

- If on the day of *Amavasya* of month of *Vaishakh* the bull sits facing north with its front leg straight in the front and chews the cud in the morning it is very auspicious. On the other hand if a bull or cow moos early morning on the day of *Akha* Teej it is believed that there would be a famine.

- If a deer is seen walking from left to right it is good but if it walks from right to left it is bad. If a deer is seen on right it is good but if seen on the left it is bad.

- If a crow comes to one's house and crows on the day one has to go to the field it is good. If it crows in the night it indicates famine. If a crow drops a bread piece in the field it is good and if it picks bread piece from the field and flies away it is bad.

- *Sugan* are also seen with the help of ants. People make a *Lata* (threshing ground) on the day of *Akha Teej* and spread some grains on it. If an ant picks up a grain and goes away it is bad. But if an ant brings a grain into the *Lata* it is good. The crop of the grain which is brought by the ant is supposed to be very good. If, for example, an ant brings a grain of Bajra then it is thought that the crop of Bajra would be good.

- If going to field a married woman is seen coming with a pot filled with water it is good. If the pot is empty it is bad. If one sees a barren woman or a widow or a woman with her hair loose while going to the field it indicates famine.
- If in the month of *Savan* the wind blows from the east it indicates rain.
- If the farmer gets up in the morning and sees a single cloud in the sky it is a good *Sugan* and indicates rain.

Cost

There is no cost involved in seeing *Sugan.* Usually *Sugani* do not charge any fees for telling their predictions. Usually they are ordinary farmers like others but have developed these special powers. The farmer may give some grain to the *Sugani* if the crop is good.

Effect

If a farmer relies on the *Sugani* and his prediction fails then the farmer is put to a loss. For example, if the *Sugani* predicts good rains and the farmer sows Bajra but rains fail then he is put to a loss. *Sugani* Narpat Singh of Raneri assessed his own predictions as follows:

Year	*Prediction*	*Actual*
Last year	Good crop	One-half of expected
Year Before	Good crop	Good crop
Year Before	Good crop	Good crop
Year Before	One year's grain	Half year's grain.

Festival

Sugan is seen on festival days. It is seen on the day of *Akha Teej* and *Haali Amavasya* in particular. Mondays and Fridays are preferred for seeing *Sugan.*

ANALYSIS

Being able to predict the rain is of immense importance for dry land agriculture. The main issue is that of reliability. Meteorological prediction is not area-specific. Even if rainfall is normal for Rajasthan as a whole, it may be deficient in a particular area. But no study of reliability of *Sugan* and meteorological predictions is available. *Sugan* predictions can be immensely beneficial if predictions are reliable.

SUGGESTIONS

A comparative study of *Sugan* and meteorological predictions may be made. The omens used by the *Sugan* in making the predictions may be noted and tallied with the results. It is necessary to develop this technology as a science. It would be useful to make a survey of the *Sugan* before the monsoons and then compare the actual results with their predictions.

Seeing *sugan* at lata on Akha-Teej

Sugani Kardha Ram of
Digadi

Sugani Narpat Singh of Akhadana

38

Hindola

Category: Potentially Viable

Diwali is the festival of lamps and it is celebrated for prosperity and wealth. Farmers prepare and light a special lamp on this day. This lamp is called Hindola. It is believed that crop of melons (Kachara, Ber and Matira) becomes sweet when this lamp is shown.

When is it done?

It is done in the night of *Diwali*. It is done on every *Diwali* except when there is a famine and watermelons have not grown.

Where is it done?

It is done at home.

How is it done?

A large watermelon is taken. A hole is made on the top and its flesh and seed are removed. Holes are made in the body of the melon in different shapes. A clay lamp with a cotton wick and oil is lit and placed inside it. It is taken to neighbouring houses and to cattle sheds and also into fields. While doing this, the following song is sung:

Diwaali Raa Diyaa Deetthaa

Kaachar, Bor, Matiraa Meetthaa

(When the lamp of Diwali is shown

Kachara, Bor and Matira become sweet.)

Festival

Hindola is made on the festival of *Diwali*.

Effect

Hindola is the lamp that is lit during the festival of *Diwali*. It is reported that the *Hindola* is then preserved. Pieces of the dried melon are fed to the large livestock when they suffer from the disease of *Khurad*.

ANALYSIS

All the nine families surveyed in Village Shiv, Badmer made *Hindola*.

Number of respondents who made Hindola	*Reasons*
9	*Hindola* represents peace and prosperity. The light of *Hindola* brings happiness and comorts at home. *Hindola* is also shown to cattle for prosperity of cattle. It is also believed that Kachara, Ber and Matira become sweet after a *Hindola* is shown to the crop.

SUGGESTIONS

A study may be undertaken to find out the efficacy of use of *Hindola* in checking the disease of *Khurad*. Other benefits, if any, may also be ascertained.

Hindola being made

Hindola

XIII
Water Management

39

Bund Construction (Khadin)

Category: Potentially Viable

A bund is constructed on the borders of the field so that water does not flow out. Rainwater is retained in the field. The soil then retains moisture for a longer time. The water flowing in brings with it organic matter, minerals and new soil. These are retained in the field and act as manure. Because of more moisture, even two crops can grow in a year in one field.

When is it done?

The *Khadin* is constructed after the farmer has harvested the *kharif* crop between December and June.

Where is it done?

Khadin is built on the side of the field which is lower. A bund is built to trap the water in the field.

How is it done?

Earth is dug from the field and a *bund* is made from it. The dug up area—called *Agor*—becomes a depression, which collects rainwater. A cemented structure—*Nehta*—is made to allow for excess water to overflow for otherwise the pressure of water often breaks the *khadin.* A G*hatnala* is also built or a pipe is put in the Khadin for removing the extra water if need be. This is used when the farmer wants to empty the *khadin* of water below the *Nehta*. It is opened if the crop of Bajra standing in the bed is likely to be spoilt because of water logging.

Cost

Manual Construction

The cost of making a *khadin* on a 10-acre plot of land is estimated below. The dimension of the *bund* are taken as follows:

Length	2000 ft
Height	5ft
Width (bottom)	15ft
Width (top)	2ft
Total earth to be excavated	85,000 cft.

One labour digs 100 cft. Total labour required is 850 mandays @ Rs. 60 per manday = Rs. 51,000.

The cost of *Nehta* is estimated at	Rs. 5,000.
Total cost of *Khadin* is about	Rs. 56,000.

Tractor Construction

Nowadays tractors rather than manual labour are used to make *khadin.* To make a *Khadin* 2,000 feet long and five feet high it takes 300 hours

Rent of tractor and Karaali for one hour	Rs. 150
Total expenditure for 300 hours @ Rs. 150 per hour	Rs. 45,000

Constructing Khadin by Strengthening Baad

Some farmers have made *khadin* not by earthwork but by reinforcing the *Baad* year after year. This can be done only where the low-lying part of the field lies against the windward direction. The soil is arrested by the *Baad* on the low end of the field. Gradually the baad becomes big and begins to retain water.

Chokla Ram of Raneri has built a *Khadin* of 500 *bigha* in this manner. He cultivates Bajra on 250 *bigha* and says that the yield is double of that which he would have got otherwise.

Income

The income depends upon the rains. About one-half of the area may be suitable for growing mustard. A quick estimate of net income of Rs. 2,500 per acre for (unirrigated) Mustard leads to a yearly income of Rs. 12,500. The difficulty is that this income is highly uncertain, as it is dependent on the rains.

Sometimes the pressure of water breaks the *Khadin* during the rains. In such a situation the farmer has to incur a heavy loss. He does not cultivate Bajra because water logging would spoil it. And, he does not get the *rabi* crop because the water flows away and the soil does not retain much moisture. In 2001 nearly every *Khadin* in Raneri had got broken in this manner. This may be a major disincentive for people to build *Khadin* from their own resources.

Discounting this by 50 per cent, the income is estimated at Rs. 6,250. The rate of return comes to 11 per cent for manually built *khadin* and 14 per cent for built by tractor, which is well below the opportunity cost of capital of 24 per cent plus. Hence people do not generally build *khadin* from their own resources .

Effect

The *Khadin* traps the rainwater in the field. It prevents erosion of the soil by water. The organic manure and minerals from the field do not flow away with rainwater. Plants and agriculture waste lying in the field also get converted into manure due to the stagnating water. The farmer is generally assured of at least one good crop. The yield of Bajra in *Khadin* is estimated to be double that of on a normal field. Sometimes he can get two crops from the same field. At other times, if the *Khadin* gets broken, he gets none.

ANALYSIS

The construction of *Khadin* is largely dependent on the economic factor. The rate of return of a *Khadin* constructed by undertaking earthwork is between 11 and 14 per cent, which is not economical.

SUGGESTIONS

The design of *Khadin* must be improved to minimize the loss due to breach of the *Bund*. A structural engineer, as well as a team of village technicians, may be asked to look into the ideal design and dimensions.

The construction of *Khadin* by strengthening the *Baad* by more systematic *Kanabandi* should be studied and propagated. It is a low-cost technology. The topographical conditions (slope, wind speed and direction, soil type, etc.), which make it possible to construct such *Khadin*, may be enumerated and propagated.

Khadin under construction

Nehta of Khadin

Kadin made by gradually reinforcing *Baad* at Raneri

40

Storage of Drinking Water (Tanka)

Category: Stable

There is a perpetual shortage of drinking water in the semi-arid deserts of Western Rajasthan. The only source in many areas is the rain. Special tanks called *Tanka* are built to collect and store water. The tank—typically about 10 feet deep—is built underground. The catchment area of about 30 feet radius is made sloping towards the tank inlet. Rainwater flows into the *Tanka* by gravity. The water here is stored and used for domestic purposes. Water stored in the *Tanka* generally meets the requirements of the family for six to nine months. For the rest of the period people have to collect water from *nadi* or wells, which may often be located as far as a few kilometres from the homestead. People who live on their far-off fields and do not have easy access to a well or a *nadi* build *Tanka* more frequently.

Tanka is also used to store water that may be purchased during summer months. Often people will buy a tanker full of water—say 3,000 litres. This water is stored in the *Tanka* and used for a week or ten days till it lasts.

In a few instances, the rich in the community have built *Tanka* as an act of charity.

One modification done is to collect rainwater from the roof and take it through pipes to the *Tanka*.

When it is built?

The Tanka is built during the winter and summer seasons when the demand for labour in agriculture is less. The water is used during and after the rains for as long as it lasts.

Where is it built?

Tanka is built near the homestead.

How is it built?

A pit is dug in the earth. It is lined with stones cemented together with lime mortar. A catchment area is built around the tank sloping towards its inlet. The catchment area is hardened by pounding or by laying a layer of *kankar* or other hard construction material. A small pit is made at the inlet, which works as a sand trap.

Cost

The cost of construction of a *Tanka* is about Rs. 10,000. The break-up is as follows:

(in Rs.)

Digging	600
Cost of stone slabs	1500
Lime for mortar	2750
Transport	350
Bajri (hard earth)	750
Labour catchment	500
Labour stonework	3100
Labour water during construction	150
Door on *Tanka*	400
Total	10,100

It will be seen that much of the cost is in the form of labour. Often people will cut the stone or collect the *bajri* themselves.

In the absence of a *Tanka* one woman of the family would have to work for about one-half day to fetch water. At a wage rate of Rs. 40 per day, it works out to Rs. 20 per half day or about Rs. 7,000 per year.

It is highly economical to build a *Tanka*. By incurring a capital expenditure of Rs. 10,000 one can save about Rs. 7,000 per year. Of course, this does not include the opportunity cost of land.

Effect of the Tanka

Availability of *Tanka* implies that women would not have to leave home to fetch water. They would be able to stay at home and take care of the children and the men would be able to go out for wage labour..

Tanka uses up agricultural land.

The family gets an assured supply of water for drinking, bathing and other household uses for many months. Women do not have to walk long distances to bring water.

The problem of untouchability still persists in many villages. Upper-caste persons may not allow the lower-caste people to take water from their wells. Building a *Tanka* saves the lower-caste people of the consequent humiliation.

Much money would be spent in purchasing water in the absence of a *Tanka*. Or women of the family would have to walk long distances and carry water.

Tanka makes available drinking water for animals and leads to greater development of livestock.

Having a *Tanka* is considered to be a measure of social respect.

Festivals

Construction of a *Tanka* is commenced on an auspicious day and time.

Roof Top Water Harvesting

Instead of making catchment area around the tank, nowadays pipes from roofs tops of houses are linked to the tank so that rainwater flows through the pipes into the tanks.

How is it done?

The Tanka is constructed as mentioned above. Rain water pipes from rooftop are placed to bring rainwater into the *Tanka*. This is possible because nowadays *pucca* houses are being built. This option was not available earlier when *kutchha* houses were being built.

An additional cost of Rs 500-1000 is involved in buying the pipes and placing them. The cost is dependent upon the design of the house and distance of the *Tanka*.

Effect

Land, which was earlier used for catchment area, can be used for other purposes.

ANALYSIS

There is no modern alternative to the *Tanka* as far as people living on dispersed homesteads located in the midst of their fields are concerned. The only alternative is to fetch water through tractor-tankers and store water in smaller tanks. This is prohibitively expensive. Therefore the technology of the *Tanka* is stable for field-based homestead.

The Public Health and Engineering Department in the settled villages is often supplying piped water. People also draw water from village ponds and *nadis*. *Tanka* is not appropriate for such conditions where sources of running water are available.

There is no dispute regarding the efficacy of roof top water harvesting. A survey shows that about one in five persons is already using this technology.

Survey

Name of Technique	*Number of respondents using it*	*Reasons*
Traditional : Tanka with catchment	17	*Tanka* provides water at home. Women have enough time as they do not have to fetch water.
Traditional:Tanka with roof top water harvesting	4	Land, which was used for catchment area, can be used for other purposes. Catchment area has to be repaired after some time. The roofs are cemented and do not have to be repaired.

SUGGESTIONS

1. It will be advisable to have water samples from *Tanka* checked for suitability as potable water. It is possible that some diseases may be occurring due to use of water from *Tanka*. If so, measures for disinfecting the *Tanka* should be developed and awareness of the same may be spread.

2. Water harvesting from rooftops should be added to the *Tanka* construction. Previously *pucca* houses were few. Perhaps pipes were

expensive. Thus rooftop water harvesting was not much practiced except in areas like Shekhavati where ground water was brackish and unfit for potable use. Now this dimension can be added to the *Tanka* construction. Demonstration can be made in villages and publicity material can be distributed. This will save valuable land from being used as catchment area.

Tanka under construction

Sand trap of tanka

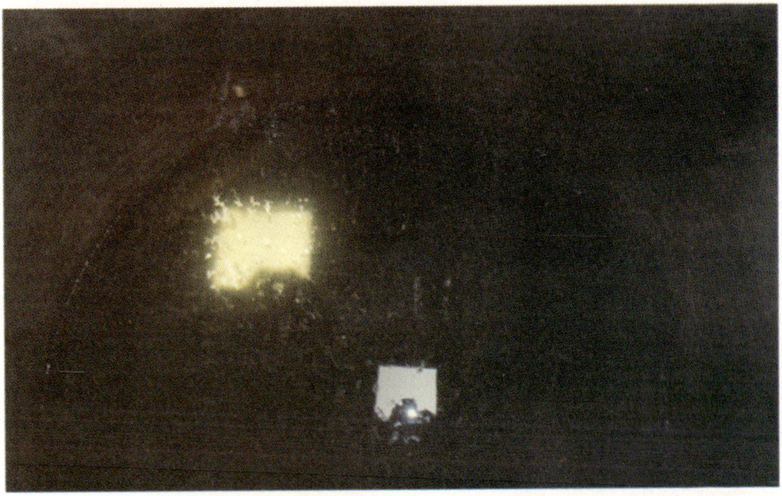

Inside of tanka

Completed tanka

Completed tanka

A community tanka in gochar

41

Water Tank (Johad)

Category: Potentially Viable

Johad is a water tank meant to provide drinking water for cattle as well as humans. It is generally made on common public land so that all members of the society have equal right to use it. The land where it is must be level and must have clay soil so that rainwater could collect there. Every year the tank has to be cleaned.

When is it made?

Johad is made in the lean period between January and June. It is cleaned every year in the months of April to June when there is less water.

Where is it done?

Most *Johad* are made on the peripheries of villages. A *Johad* must not be very far so that people could easily access it and get water from it.

How is it done?

Most *Johad* are about 100 ft. long and equally wide with a depth of up to 10ft. The site is selected in a depression so that water from the catchment around it is collected. The *Johad* gets silted gradually. The silt has to be removed every year or after a few years as the need may be. All villagers do this work together. Two or three members from every family get together and remove clay from the tank's bed for a few days each year. If a family does not participate in the work, it is expected to pay something for labour.

Cost

Cost of the *Johad* depends upon its dimensions. New *Johad* are rarely being made nowadays; hence it has not been possible to assess its cost as also the cost of the *Nadi*.

Effect

Rainwater gets stored in the *Johad.* It provides drinking water for livestock in particular. Women have to bring water for the family and the cattle. They are saved the trouble of walking miles to get water from some far-off source.

Festival

On the day of *Teej* in the month of *Savan* unmarried girls of the village take bath in the *Johad* and pray to Lord Indra to keep the tank ever filled. It is believed that this ritual keeps the tank ever overflowing with water.

ANALYSIS

Johad is made where the livestock go for grazing. Increasingly grazing lands are being cultivated hence the utility of this technology is declining. There is a strong tendency to sell cows and bullocks and replace them with buffaloes, which do not require to be grazed. This reduces the need of the *Johad.*

SUGGESTIONS

- Buffalo is a water-loving animal. A study may be made of the impact of the availability of *Johad* on the economics of the buffalo. If the impact is positive and significant then a programme to make *Johad* close to the village on common lands may be undertaken.
- A social consensus may be built in the village on the need to collectively desilt the *Johad.*

Dry johad

Johad

42
Village Tank (Nadi)

Category: potentially Viable

Nadi is a water body that stores rainwater. All people in the village have access to it and use its water when there is shortage. Sand also flows into the *Nadi* with rainwater. It settles in the bottom and reduces the capacity of the *Nadi* and also makes the water muddy. The sand is removed from the *Nadi* every few years so that water remains clear. Removing clay from the bottom increases the water retention capacity of the *Nadi*. The sand and soil removed from the bottom is clayish and used to build *kutchha* houses.

CONSTRUCTING A NADI

When is it used?

A *Nadi* is built when the village is established. In fact, villages were established near areas that could easily be made into a *Nadi*. The *Nadi* used to be the only source of drinking water. Often the zamindars or Kings or village folk themselves would construct a *Nadi*. Most of the *Nadi* seen now are very old. A Rajput king in the memory of his wife Sunderi reportedly constructed the *Nadi* at Akhadana five hundred years ago.

Where is it made?

A *Nadi* is constructed near the village.

How is it done?

A bund-like structure is made on the lower side of the site to increase the capacity of the *Nadi*. A *pucca* overflow is constructed to allow excess water to flow away. A mound is made with the earth that is excavated. This is called *Lakhota*. The *Lakhota* appears to have no economic utility.

However, it is visible from distance and could have helped people easily locate the *Nadi*.

The catchment area is kept clean of animals lest dung flow in with rainwater and create fermentation in the *Nadi*.

Cost

All the *Nadi* existing today have been constructed decades ago. No estimates for their cost are available. The cost depends upon the topographical features. People estimate that it would take about 10,000 man- days to construct the *Nadi* at Akhadana. At the present wage rate of Rs. 60, the cost would be about Rs. 600,000.

Effect

Nadi is the lifeline of the village. There could have been no village without a *Nadi*. It was *the* source of water in the past. Ground water in the area is low and could not be drawn with the traditional technologies then known. The only other source of water is the *Tanka*. But that only provides limited drinking water. Its catchment area is very small—about 100 square feet only against 10-20 acres of the Nadi.

Cleaning a Nadi (Traditionally)

The *Nadi* has to be desilted periodically to maintain its capacity as well as cleanliness.

When is it done?

Cleaning of *Nadi* is done when there is none or very less water in the *Nadi*. It is done between March and June. It is especially done on days of *Ekadashi*, *Poornima* and *Amavasya* in the months of *Baisakh*, *Ashad* and *Jyeshtha*—broadly conforming to the same period. Village talk is that women removed the mud on these auspicious days after completing their fasts. But none in a group of about 20 women at Akhadana was willing to say that she had indeed done so.

Where is it done?

It is done in the bed of the *Nadi*.

How is it done?

In the olden times, on the days of *Ekadashi*, *Amavasya* and *Poornima* women used to go to the *Nadi* and remove clay from the bottom. This was done with a religious fervour and as a duty. Garbage was also removed from the bottom. The area around the *Nadi* was also cleaned and garbage and dung was removed and burnt.

At other times people from the village would get together on the above-mentioned days and remove clay from the bottom. Or villagers would collect funds and get the cleaning done through labour. The clay removed from the *Nadi* would be used to reinforce the bund on the lower end to increase the capacity of the *Nadi.*

The main obstruction to the use of *Nadi* appears to be that of management. The water remains clear only if animals do not enter the *Nadi* and even the catchment area and if people do not take bath in the *Nadi.* Enforcement of these rules requires not only a social consensus but also punitive measures against the transgressors. These are not possible in the present period. Chokla Ram of Raneri was asked why was it that the people in Akhadana maintained their *Nadi* but people in Raneri did not. He replied said, "After the Public Health and Engineering Department (PHED) water had become available people have lost interest in the Nadi. Rules exist but they are not followed. In Akhadana, there is no PHED supply at Akhadana hence people still maintain the rules. Because water is available from PHED hence the elders do not take interest in enforcing the rules. They have stopped giving attention to this matter." In this village the PHED supply had failed for ten days two years ago. Then there was a sudden revival of interest in maintaining the *Nadi.* The following rules were once again made and enforced:

1. The tire of an animal cart will not touch the water.
2. Animals will be removed immediately after they have drunk water.
3. No one will take a bath.

These rules had then once again become dormant after PHED supply resumed. This clearly shows that people prefer PHED supply but will look after the *Nadi* if such better alternative is not available.

The rules for use of *Nadi* at Akhadana were as follows:

1. No one will excrete in the Nadi.
2. No one will wash his feet or take a bath.

If anyone violated these rules he would be fined Rs 151 and if he still persisted then he would be ostracized from the community. But Babu Singh of Akhadana said, "Previously people who violated the rule were punished. Now due to party politics in the Panchayat there is no unity in the village. The village elders do not have the authority to recover such fines. As a result the *Nadi* is not being maintained." Clearly, the *Nadi* is a part of the traditional socio-political system of local governance. It is perhaps not possible to manage a *Nadi* in the modern court-and magistrate-based governance.

Cost

The cost of cleaning a *Nadi* depends upon its size and the distance involved in carrying the earth.

Effect

The water that gets collected in the *Nadi* is clean. It increases the water holding capacity of the *Nadi*. Water is retained for a longer time. People get water easily and they do not have to fetch water from long distances through tankers from other tanks or *Nadi*. People get clean water to drink and are healthier.

Guman Singh of Jelu said, "People were not permitted to wash hand and feet in the *Nadi* previously. Now because the tubewell is available people are taking less care of the *Nadi*. Cattle are regularly entering the *Nadi*." Generally people prefer water from *Nadi* to that of tubewell (please see survey below). This preference for the tubewell over *Nadi* may be because of the problems of management of *Nadi*.

The future of the *Nadi* depends upon the stability of supply of drinking water by PHED. Instances are reported where the people initially ignored the *Nadi* when drinking water became available from tubewells. But soon the water from tubewells became brackish and people reverted back to maintaining their *Nadi*. This is reported to have happened at Shekhasar. A similar situation developed in Tekra where the tubewells went dry after a few years. At Malam Singh-ki-Seed the PHED supply failed after four or five years and people reverted to their *Nadi*.

Festival

Nadi is, theoretically, cleaned on the days of *Amavasya* when people do not do any other work and are free.

On the day of *Nirjala Gyaras*, the eleventh day of bright fortnight of the month of Jyeshtha, villagers keep fast and clean the *Nadi* as a religious duty. The women sing hymns.

In the month of *Shravan*, on the day of *Teej*, the women keep fast, wear jewelry and go to the *Nadi* and perform worship. They set afloat leaves of Aak on the surface of the water. After cleaning of the *Nadi* the women eat food and end the fast.

On the day of *Akha Teej* too women would bring water from the *Nadi* and cook *Khichadi* for the whole family.

Cleaning a Nadi (Modern)

At present people are not interested in cleaning the *Nadi*. Some government or private agency does the cleaning with the help of tractors. The community feeling is lacking and no longer do people get together to clean the *Nadi*. People are more willing to contribute funds rather than their own labour. Many people go out of the village for work and are unable to contribute their labour. However potters still continue to remove the clay from the *Nadi* bed to make pots. The *Nadi* at Akhadana was cleaned by such efforts on the following dates:

1980	By village people collectively by self-effort
1991	By the government
1995	By NGO (Gravis) on 50% subsidy
2000	By NGO (Gravis) on 50% subsidy
2001	By the government under famine relief work (please see figs)

It appears that the self-effort of the villages for cleaning the *Nadi* has been lost in the recent period.

Effect

The community feeling is becoming less and no longer are people ready to offer selfless service. The responsibility of taking care of village resources is shifting to the government. The *Nadi* is not cleaned on time or at regular intervals. It depends entirely upon the schemes such as food-for-work launched by the government or NGOs getting grants to have the work done.

Tubewells

Nowadays people get drinking water from tubewells installed for the purpose of irrigation.

Effect

Water is available easily. It becomes easier for women to get water. Even neighbours can get water. Generally, tubewell owners do not prohibit people from using water for drinking and home purposes.

The tubewell water is sometimes sweet and other times brackish. People use water from those tubewells that have sweet water.

People often prefer water from tubewells to that of a *Nadi*. It is clean, not so brackish and easily available. Water from *Nadi* may not be clean particularly because the rules for keeping the catchment area clean are not followed these days.

Cost

Setting up a tubewell can cost up to Rs 3-4 lacs. But this is mainly for irrigation.

PIPED DRINKING WATER (PUBLIC SUPPLY)

The Public Health and Engineering Department (PHED) has the responsibility to ensure piped drinking water reaches all the villages of Rajasthan. This water is supplied free at the stand posts. Usually pipe connections are not provided to individual houses.

When is it provided?

PHED supplies piped drinking water as much as possible throughout the year.

Where is it supplied?

It is supplied at a convenient place in or near the village.

How is it done?

Water is extracted from the ground where available or tapped from the supply from Indira Gandhi Canal. It is then clarified and supplied by pipes to the village tanks. Mixing with sweet water in optimum quantity reduces the brackishness. A tap is installed for village people to collect water. A trough is made to enable animals to drink water. No charge is levied upon the people for this water.

Cost

The cost of water is borne by the government.

Effect

Water of reasonable quality (i.e., not much brackish) is available easily to the village people. It is generally of good quality from the point of view of health.

ANALYSIS

A comparison of the three sources of drinking water is made in the table below. Points are given on a scale of 3 for arriving at a conclusion.

It will be seen that PHED supply is ranked first with *Nadi* second and the tubewell third. The *Nadi* has the advantage that it provides sweet water close to the village. This indicates that people will be interested in maintaining the *Nadi* wherever PHED water is not available. Babu Singh of Akhadana—one village where the *Nadi* was still being maintained—said that "people will forget the *Nadi* if water from PHED was available."

Source of water	*Quality*	*Cost*	*Distance*	*Regularity of supply*	*Touch earned*
Nadi	(2½) Sweet But uncertain from health point of view	(2) Medium Although no direct cost is incurred in drawing water, the system works only if people contribute to the cleaning of the *Nadi*. This must be treated as a cost.	(3) *Nadi* is located near the village	(2) Medium. Supply generally lasts till early summer.	(9½)
Tubewell	(2) Medium. Water is often medium brackish.	(3) Low. Drinking water is supplied free from tube wells.	(1) Tube-wells are located far in the fields. Cost of transport is involved.	(3) Water is available throughout the year. (But only at time when power supply is available. This is a minor matter	(9)
PHED	(2) Medium. Water is often medium brackish.	(3) Water is supplied free of cost.	(3) Supply is made near the village.	(2) Depends upon location. Some villages located near supply centres get regular supply. Other located at tail end get supply once in ten days or less.	(10)

Note: Figures in parenthesis are points allotted.

This is in consonance with the results of a survey. People prefer *Nadi* water to that of tubewell. Survey of PHED was not done hence no inference can be drawn on that count.

Survey

Name of Technique	*Number of respondents using it*	*Reasons*
Traditional: Nadi	33	*Nadi* water is sweet and good while that from tubewell is heavy. Clay from *Nadi* bed can be used to make earthen pots and vessels.
Modern: Tubewells	3	*Nadi* water is not clean. It causes diarrhoea and cold.
Modern: PHED	Not Surveyed	NA

Evidence of the preference for the PHED water is that the *Nadi* at the town of Phalodi has completely dried up because PHED water is regularly available.

It would be unwise to abandon the *Nadi* before the supply of sweet drinking water from tubewells or PHED is assured in a *sustainable* manner. The efforts to revive the *Nadi* should be continued for the present.

SUGGESTIONS

A method has to be found to enforce the rules for the use of *Nadi*. The Panchayat elections have disabled the traditional consensus-based approach. A window has to be created within the rules of Panchayat to manage local community resources collectively. Linking the rules of *Nadi* to religion and festivals can increase the participation of people.

Akhadana nadi

Lakhota of Akhadana nadi

Jelu nadi

Cleaning of nadi in progress

PHED supply at Raneri

PHED supply at Raneri

Community well at Kinsaria

43

Drawing Water from The Tank Bed (Beri)

Category: Uncertain

The water from the *Nadi* does not often last through the summer. Small wells—called *Beri*—are dug in the bed of the *Nadi* to extract drinking water in such circumstances.

When is Beri used?

Beri is used during the summer months when the *Nadi* has become dry.

Where is it made?

Beri is made in the bed of the *Nadi* where highest concentration of ground water is likely to be found.

How is it done?

A *kutchha* well of about 4-5 feet diameter is dung in the bed of the *Nadi* when it is dry. Water is extracted as per the requirement. The mouth of the well is closed by stones when the rains come and the well remains submerged in the *Nadi* till the next summer. It is reopened when the need arises again.

There is a good chance of the Beri collapsing when it is submerged in water. The *Beri* is refilled with sand and then the mouth is closed in the first one or two years. The earth is again dug out in the next summer. The *Beri* becomes pucca after two or three years of such operation. The life of the *Beri* after it so stabilizes is estimated to be about 100 years. The *Beri* is treated as the personal property of the owner though the *Nadi* itself generally belongs to the community.

A *Beri* provides about 25-50 buckets or about 750 litres of water in a day.

Cost

It takes about 3 persons 25 days to construct a *Beri*. At a wage rate of Rs 80 per day for one person this works out to Rs 6,000. The expenditure in filling the *Beri* with sand and removing the same for next two years may be taken to be about the same amount. The total cost may be about Rs. 10,000 to Rs. 12,000.

There is little cost of maintenance after the *Beri* becomes stable.

Effect

The Beri provides valuable drinking water when the *Nadi* is dry. It is the source of life during the summers.

ANALYSIS

The cost of water—one tank of 750 litres per day for 75 days—from a Beri is estimated as follows.

Fixed Cost	
Interest on Rs 10,000 @ 24%	Rs. 2,400
Depreciation Rs 10,000 at 100 years	Rs. 100
Variable Cost	
Cost of lifting water	
75 days x 2 persons x2 hour x Rs 80/8 hours	Rs. 3,000
Total	Rs. 5,500
Per Unit East	
Cost per kilolitre (75 days x 750 litres)	Rs. 98
Variable cost per kilolitre	Rs. 53

The cost of water from a tractor tanker, which supplies from a tubewell, is Rs 250 for 6,000 litres or Rs 42 per kilolitre.

The tractor-tractor combination is cheaper. The *Beri* also involves an initial investment of labour worth Rs 10,000. Therefore the Beri is slowly slipping into disuse.

It should be remembered that this survey covers only those who own a *Beri*. A large number of others are getting water from Tractor-Tubewells. Two among the seven owners are entirely getting water from Tactor-Tubewells. Others are using only a few of their *Beri*. The reason may be that many family members have shifted to the city and the requirement of water for the family is less than earlier.

A survey of Beri in the Akhadana Nadi reveled the following.

Name	*Number of Beri owned*	*Number of Beri in use in 2001*	*Comments*
Bhanwar Singh	3	1	One is sufficient. Will open other Beri if need arises.
Babu Singh	5	2	Ditto
Bhoom Singh	1	Closed	Get water from Tractor-Tubewell at Raneri
Om Prakash	2	1	Ditto
Ramu Ram (SC)	1	Closed	Takes water from *Beri* of another.
Bhura Ram	1	1	Using it
Aru Ram	2	1	One is sufficient

Yet, five out of seven persons who own *Beri* do draw water from the same. The reason is that the fixed cost of *Beri* having already been incurred, it is cheaper to obtain water from *Beri* than from the Tractor-Tubewell. The variable cost of drawing water from the *Beri* is Rs 53 per kilolitre against Rs 42 from Tractor-Tubewell. If the opportunity cost of labour is less then it is more economical to draw water from Beri.

No new *Beri* has been constructed at Akhadana in recent years.

The future of *Beri* is linked to that of the *Nadi*. If people have to rely on the *Nadi* for sweet drinking water then the *Beri* too will be revived because the water supply during summer months from the *Nadi* is possible only through the *Beri*.

SUGGESTIONS

Try modern methods of constructing the *Beri* such as by the use of prefabricated concrete rings.

Mouth of beri

Inside of beri

Drawing water from beri

Closed mouth of beri

44

Transport of Drinking Water (Kalsi)

Category: Declining

Farmers use the water collected in their *Tanka* during the rains while it lasts. Thereafter the source of drinking water in the village was the *Nadi* or the village tank. People drew water from the Nadi and transported it to their houses in large earthen pots called *Kalsi* or leather containers called *Pakhal*. The latter, however, has virtually disappeared from the village scene at present.

When is the Kalsi?

The *Kalsi* is used after one's own *Tanka* has become empty. Generally it is used in the summer months beginning in March up to June.

How is it used?

The *Kalsi* is placed upon the cart. Water is filled into the *Kalsi* at the *Nadi* and then it is transported to the homestead. The *Kalsi* is downloaded from the cart and water is drawn from it as per requirement. *Pakhal*, on the other hand, was tied on the camel or donkey's back.

Effect

Women are saved the trouble of carrying water from far-off places in pots carried on head.

Water level in the *Nadi* goes down and, as a result, stray animals are deprived of water.

Cost

The cost of a *Kalsi* is about Rs 80 excluding the cost of the animal and the cart.

The cost of a *Pakhal* is estimated to be around Rs 500. However, the *Pakhal* is tied directly on the animal's back and there is no need to acquire a cart.

The cost of one *Pakhal* or *Kalsi* of water is about Rs. 30. A cart owner can supply about three to four loads of water in a day and earn about Rs. 90 to 120.

Festival

Carrying empty containers is considered inauspicious hence water is not fetched on festival days.

METALCONTAINERS

Metal containers are gradually replacing the *Kalsi*. Metal tanks of about 3,000-litre capacity are used to transport water using animal carts. These are referred to as *Chhakada* though the word literally means the animal cart. Large tankers of about 5,000-7,000-litre capacity are used with tractors.

A camel cart can deliver up to four tanks of water in a day.

ANALYSIS

The *Pakhal* costs about Rs 500 while the Kalsi costs only Rs 80 for the same quantity of water. However, additional investment is required in a cart. It appears that most farmers already own a cart, as it is useful for transporting various goods to and from the field. Hence Pakhal is no longer being used.

The cost per litre of water from the *Kalsi*, *Chhakada* and tractor tanker has been estimated in the table below.

Mode of Transport	Capacity	Price	Cost per 1000 *litres*
Kalsi	100 litres	Rs 30	Rs 375
Chhakada	2000 litres	Rs 50	Rs 25
Tractor Tanker	6000 litres	Rs 250	Rs 42

It will be seen that the *Kalsi* is very expensive in comparison to the *Chhakada*. However, the *Chhakada* involves a cash outgo for not everyone owns a *Chhakada*. Those farmers who have no other employment may therefore prefer to use a *Kalsi* to fetch water rather than pay cash for the *Chhakada*.

The cost from tractor is higher than that from *Chhakada*. But the tractor can transport water from long distances of up to 15-20 kilometres. Therefore the tractor is used for long distances while *Chhakada* is used for short distances.

The increasing use of metal containers is confirmed by a survey.

Survey

Technique name	*Number of respondents using*	*Reasons*
Pakhal or *Kalsi*	5 (Only one person explicitly mentioned use of *Pakhal*)	It is cheaper. It makes use of animals that are otherwise idle.
Chhakada or Tractor	13	A tanker lasts several days. More water can be brought in less time. It saves time.

The difference of cost between *Kalsi-Pakhal* on the one hand and *Chhakada*-Tractor on the other is about 10 times. It is too glaring to be ignored. No advantage, such as the quality of water etc, is mentioned in respect to the traditional technologies. It seems that only such farmers are using *Kalsi-Pakhal* who have animals and no cash income. The use of this technology is, therefore, predicated on poverty.

SUGGESTIONS

The cost of a *Pakhal* could possibly be much lowered by the use of materials such as plastics or cloth. It is reported that the *Pakhal* is being made of *Tripal*. The poor farmers who make do with little water in absence of cash income may continue to use this technology.

Transporting water in kalsi

Using water from kalsi

· Index